Automation: The Future of Weed Control in Cropping Systems

Stephen L. Young • Francis J. Pierce
Editors

Automation: The Future of Weed Control in Cropping Systems

Springer

Editors
Stephen L. Young
Department of Agronomy and Horticulture
West Central Research & Extension Center
University of Nebraska-Lincoln
North Platte, NE, USA

Francis J. Pierce
Departments of Crop and Soil Sciences and
Biological Systems Engineering
Washington State University
Prosser, Washington, USA

ISBN 978-94-007-7511-4 ISBN 978-94-007-7512-1 (eBook)
DOI 10.1007/978-94-007-7512-1
Springer Dordrecht Heidelberg New York London

Library of Congress Control Number: 2013954786

Printed on acid-free paper

Springer is part of Springer Science+Business Media (www.springer.com)

Foreword

Since the dawn of agriculture, man has struggled with controlling species competitive to their crops. Many technologies and strange machines have been developed to kill or retard weeds that have often used lots of cheap energy. These days energy is not so cheap, and the indiscriminate distribution of off-target chemicals have caused widespread problems to economics and the environment as well as a detrimental public perception.

The main change in recent years has been the realization that it is very difficult to create the ability to distinguish between crop and weed species by embedding it in a chemical formulation, where it is lethal to the weed and benign to the crop and environment. These days we have a different approach by embedding the "smarts" into the applicator. It is much easier (but still not trivial) to recognize weeds from crops and carry out an "intelligently targeted input" of chemicals or energy only onto the weeds, thus significantly reducing the inputs by, in one case, 99.9 % by volume when using spray microdots. This can be reduced to zero chemicals when we use physical weeding.

Our aim in developing these technologies is to work out the minimum amount of energy that is needed to control weeds and, by extension, the minimum amount of energy that we introduce to the natural environment to turn it into production agriculture. We are at last glimpsing what I suggest is tertiary development where these technologies actually use the smallest amount of energy to kill a weed. The first example of this is using machine vision to identify the meristems of weeds and then steer a low-power laser to heat up the meristem until the cell walls rupture, thus forcing the weed to become dormant.

It might seem like science fiction, but the day where robots and smart machines are working 24/7 in our fields tending our crops is not that far off. This book goes a long way in describing the reality of weed control and the future of farming.

Head of Agricultural Engineering
Harper Adams University
Newport, Shropshire, England

Professor Simon Blackmore

Preface

In both conventional and organic cropping systems, there is an immediate need to apply the latest technologies to improve the efficiency and economics of management while reducing the impacts. Never before has there been such pressure on farmers globally to produce more with less and reduce inputs in cropping systems. The main purpose of this book is to provide the current state of automation for weed control in cropping systems, which demonstrate how being more precise in our applications is possible now and into the future.

By bringing together biologists and engineers working in the same field, ideas can be stimulated on ways to change current weed management for the better. After many discussions, some lengthy, between the editors back in early 2009, a group was formed and a paper was written. The former met at Washington State University's Center for Precision Agricultural Systems and was composed of engineers, biologists, industry representatives, and local growers with one goal: strategize new ways to use automation to control weeds in organic vegetable crop production systems. I wrote the paper that was published in *Weed Science* on the use of automation for weed control in organic cropping systems, which was also a call for broader participation by the weed science community.

As a group, we wrote a proposal for the USDA SCRI program. At the same time, I organized a symposium for the 2010 Weed Science Society of America annual meetings on automation and machine guidance systems for weed control in cropping systems. The speakers presented information and research on current weed management techniques, automated mechanical and chemical weed control, market readiness of robotics and automated weed control, and international advancements in automation for weed control. The interest from the audience was limited to a small but enthusiastic group of weed scientists. Obviously, more awareness and education on this topic was needed on a wider context, which is what solidified my interest in the topic of automated weed control and fuelled my passion for completing this book.

From the group that met back in 2009, the speakers at the 2010 symposium, several other respected individuals, and my colleague, Dr. Fran Pierce, we have assembled the first book focused solely on automation and weed control in cropping systems. The main themes of this book are (1) weeds in conventional and organic production systems, (2) advancements in technology and current weed control practices, (3) applications of automated weed control in cropping systems, (4) economics of organic and conventional production systems, and (5) global trends and future directions for automated weed control. The objectives are to (1) provide the first complete resource on automation and robotic weed control in conventional and organic cropping systems for the student, researcher, and grower, (2) shift the paradigm that precision technology and cropping systems cannot fit into a single, streamlined production system, and (3) stimulate thoughts and ideas for broader application of new engineering solutions to traditional agricultural-based problems, such as weed control.

The production of a book of this magnitude could have profound impacts on current and future cropping systems across the globe. To date, no other resource exists on this important and rapidly advancing topic of automated weed control. In the near future, a new approach will be needed for managing weed pests, especially with the challenges of weed resistance to herbicides; off-site movement of soil, fertilizers, and chemicals; an increasingly non-agrarian public; labor shortages; economies in recession; and the continued rural to suburban land use conversion.

Stephen L. Young, Weed Ecologist
Assistant Professor
University of Nebraska-Lincoln
North Platte, NE, USA

Acknowledgments

I would like to express my gratitude to Fran Pierce who provided support; talked things over; read, wrote, and offered comments; and assisted in the production of the book.

I would like to thank the authors and coauthors for making time to put their thoughts down on paper and then revising them according to a shared vision. Their expertise is what has allowed this book to become a reality.

Thanks to Maryse Elliot, my publisher, for approaching me with the idea of writing the book.

Last but not least, I thank my wife for supporting me even during the many sleep-deprived days spent working after hours to complete what seemed to be at the time an insurmountable task.

Contents

Chapter 1
Introduction: Scope of the Problem—Rising Costs and Demand for Environmental Safety for Weed Control

Stephen L. Young, Francis J. Pierce, and Pete Nowak

Abstract Many organic and conventional producers rank weed control as their number one production cost. For organic producers particularly, weed control has become increasingly important as organic production has increased its market share. In conventional systems, herbicide resistance, off-target movement, and increased regulations have left many growers with few alternatives. Added to this is an increasing demand from the public for a safer and more sustainable supply of food. This chapter addresses the problems of mechanized agricultural systems to set the stage for the introduction and adoption of more advanced technology to meet the needs of growers and satisfy the desires of consumers.

1 Timeless Weeds

> *Autonomous robotic weed control systems hold promise toward the automation of one of agriculture's few remaining unmechanized and drudging tasks, hand weed control. Robotic technology may also provide a means of reducing agriculture's current dependency on herbicides, improving its sustainability and reducing its environmental impact.*
> *Slaughter et al. (2008)*

S.L. Young (✉)
Department of Agronomy and Horticulture, West Central Research & Extension Center, University of Nebraska-Lincoln, 402 West State Farm Road, North Platte, NE 69101, USA
e-mail: steve.young@unl.edu

F.J. Pierce
Departments of Crop and Soil Sciences and Biological Systems Engineering, Washington State University, Irrigated Agriculture Research & Extension Center, 24106 North Bunn Road, Prosser, WA 99350, USA
e-mail: fran.pierce@gmail.com

P. Nowak
Nelson Institute for Environmental Studies, University of Wisconsin, 40 Science Hall, Madison, WI 53706, USA
e-mail: pnowak@wisc.edu

S.L. Young and F.J. Pierce (eds.), *Automation: The Future of Weed Control in Cropping Systems*, DOI 10.1007/978-94-007-7512-1_1,

While biblical Adam was promised thorns and thistles as part of his punishment (Genesis 3:18), Timmons (1970) review states that few agricultural leaders or farmers became interested in weeds as a problem until about 1200 A.D. One can correctly imagine, however, that from the development of primitive forms of agriculture, weeds have presented a formidable challenge for food, feed, and fiber production. Our ancestors recognized weeds as limiters of desirable plants, sources of health problems, and degraders of aesthetics over a broad range of environments. But what are weeds? Weeds are most simply defined as "[a] plants out of place." A more poetic description was provided by Ralph Waldo Emerson who declared that "a weed is a plant whose virtues have not yet been discovered." Indeed, the ongoing search for genetic materials from plants that may prove to be beneficial confirms the need for a flexible perspective in managing those plants we call weeds.

2 The Number One Pest Problem

In both early and modern agriculture, weeds clearly rank as the primary pest problem. Today, weeds plague even the most advanced and progressive farming operations regardless of their management approach, whether organic, conventional, or sustainable. Holm and Johnson (2009) state that "throughout the history of agriculture, more time, energy and money have been devoted to weed control than to any other agricultural activity." In the USA, the vast majority of crop acres are treated with herbicides (Gianessi and Reigner 2007) accounting for about two-thirds of the pesticide expenditures for US farmers in the late 1990s (Donaldson et al. 2002). Today, the development of herbicide-resistant weeds is the major concern for farmers relying on chemical weed control, while in organic production systems, the cost and effectiveness of hand removal of weeds is a concern due to expenses, labor availability, and, in large-scale systems, the social acceptability of employing large numbers of migrant labor. Farmers are increasingly facing environment and economic consequences of emerging weed management challenges, restrictions on the availability and effectiveness of chemicals, changing government policies, and dynamic markets that can reward or punish depending on how weeds are managed.

There is no immunity to weeds and the problems they cause, whether for a large farmer or a typical home gardener. Without continued and focused management and control efforts, a low or an apparent nonexistent weed population can very quickly get out of hand with direct (e.g., lower yields) and lasting (e.g., soil weed seed bank) effects. Because weed impacts are significant and have been passed on through countless generations, there is a continually evolving array of the types and numbers of different approaches for controlling weeds. In commercial cropping systems these options are vast and include the categories of mechanical, chemical, biological, and cultural control.

3 Management: Then and Now

Prior to the development of herbicides, weeds were largely a management challenge that was addressed with planning and the use of high amounts of disturbance. Crop rotation was important, and whatever new ground was available was used once the "old" location had become too infested with weeds. The movement between and to new land parcels was, in itself, a type of rotation, although not what is typically practiced today.

Early day cropping systems relied on routine disturbances to reduce weed pressure. The use of cultivation was important for disrupting weed growth and could be applied in the simplest of forms. Unfortunately, early day cultivation could not be applied selectively, except in rows, and bare soil, which resulted in high amounts of erosion, was common in many fields. In the Midwest, the Dust Bowl of the early 1900s was caused by excessive tillage, as the prairie sod grasses were eliminated in favor of annual cropping systems. When Lowdermilk (1939) wrote his report on the demise of ancient civilizations due to excessive erosion, the cultivation of weeds in irrigated cropping systems was identified as a likely culprit. As noted earlier, weeds are timeless, and as we have to relearn again and again, the various forms of disturbance used to manage weeds may have significant consequences that ripple across both time and space.

With the invention of 2,4-D in the 1940s, weed control changed dramatically. The agricultural chemical revolution (i.e., the substitution of inorganic fertilizers and manufactured chemicals to replace manure, humus, and various forms of pest control) following WWII gave growers the ability to selectively manage weeds in cropping systems with chemicals designed to kill on contact or through movement within the plant. Later, new herbicides were developed that provided total, selective, or partial control of weeds, which gave growers great flexibility in managing weeds in their crops. These innovations also brought about an important change in the indigenous knowledge associated with weed management. Prior to the introduction of these chemicals, growers had to accrue a system of knowledge on multiple dimensions of weed control: what to do, when and how to do it, and what observations are needed to guide decisions. The increased ease associated with dependence on chemical control also meant less knowledge was required for managing cropping systems. Knowledge of weed ecology became less important, and a grower could focus on other important management aspects, including fertility, marketing, or crop selection.

Currently, the most relied upon techniques for controlling weeds in conventional cropping systems are the use of cultivation and herbicides. The invention of herbicide-resistant (HR) crops has allowed for a quick application of a single herbicide sprayed over the entire field to control weeds without harming the crop. The simplicity of this system has actually led to the emergence of HR weeds. The use of a single herbicide that is applied repeatedly in one season at high rates on mature weeds is a recipe for resistance, which occurs when an individual plant or population responds to intense selection pressure. In addition, growing the same crop each year and using the same weed management program only exacerbates the problem. Add these 'incorrect' management strategies together across large acreages and only time is needed for HR weeds to start appearing in grower fields, which they now have. Today HR weeds are a very significant problem, one that keeps increasing in size and scope, as we continue to fail in understanding that any new technology is a double-edged sword—there are many benefits, but mismanagement can lead to major problems.

In organic and some conventional cropping systems, the use of cultivation remains a heavily relied upon management tool for controlling weeds. The ability to systematically move through a field and physically disturb weeds has been one of the most relied upon control tools for centuries because there is no guess work and virtually all of the risk is eliminated. Large-scale operations use this tool because equipment manufacturers have created a wide range of implements appropriate for these operations. While the same range of equipment may not be available to small-scale growers, they have a greater capacity to respond to smaller or sudden changes than larger growers because they have an intimate relationship with their crops and fields. This type of knowledge or familiarity with the dynamics of weed ecology is extremely difficult at large scales, and since HR weeds are an increasing problem, scientists are looking to other forms of innovation to address this situation. One of the promising developments is automated and targeted weed control, a theme that is addressed in the remainder of this book.

4 Costs, Costs, Costs

All forms of modern-day weed control have costs associated with them. Some accrue to the grower, others to workers who may be exposed to chemicals, and still others to environment and society on the whole. Yet the lack of weed control diminishes yields and profits, thus resulting in an ongoing balance by growers to limit risk by falling somewhere between an 'insurance level' and minimal level of control that will minimize the impact of weeds. In conventional systems, the exposure to chemicals by those who have to make the applications is a safety risk that is costly in terms of health and finances. Although some cases are suspect, there are links between health problems and the application of pesticides in crop production systems. In addition, the locations where chemicals are manufactured are "no shining stars" of environmental excellence either, but the same could be said for fertilizer manufacturers and their various distribution points.

Not only are applicators and manufacturers vulnerable to the ramifications of handling toxic chemicals, but the environment itself suffers from any level of chemical application. Weeds suffer, which is desirable from a production standpoint, but it is debatable, often on a site-specific basis, as to whether yield benefits justify potential harm to humans and surrounding ecosystems. Non-HR crops suffer from misapplications and even HR crops have been debated as to whether they are completely suitable for the environment. Off-target movement (e.g., drift, runoff) of chemicals has numerous effects on animals, insects, birds, and fish,

Fig. 1.1 Organic onion field in eastern WA, USA with a hand-weeding crew. Every other pair of onion rows has already been hand weeded and cultivated (Photos courtesy of Rick Boydston, USDA-ARS)

although all chemicals face rigorous testing mandated by EPA (in the USA) prior to commercial sales. Nevertheless, this testing does not prevent an off-label application made by mistake or in the wrong circumstance. The debate surrounding the accounting for benefits and costs is not new and has been with us with the emergence of each new form of weed management. While Rachel Carlson may have been a lone voice when she issued the warning associated with the use of chemicals in her book Silent Spring, today there are hundreds of books and reports on how we have allowed HR weeds to become a major agricultural issue (Beckie 2006; Beckie et al. 2006; Beckie and Tardif 2012; Bhowmik 2010).

In organic systems, similar costs to the environment can occur if an over-reliance on cultivation is used. The continued disturbance of the soil leads to excessive erosion by means of both wind and water. Since weed control can be more difficult in these systems, it could be argued that excessive weeds that are left uncontrolled are also polluting the environment. Probably, this is one of the main reasons why there are so few large-scale commercial organic farm operations. For those companies that are successfully producing organic crops, one of their biggest inputs is manual labor, a significant economic cost to the grower, and one that challenges the notion of a sustainable system due to these social dynamics (Fig. 1.1).

The costs for weed control, other than to the environment and applicator, can range from minimal to financially devastating. In many countries, manual labor is used to control weeds because it is cheap and plentiful. Most often, in these situations, other challenges exist that relate to growing, processing, or delivery of crops to market. In locations where labor is not widely available, costs are reduced by using chemical weed control because it is relatively cheap and easy to use.

Increasingly, the environmental costs of weed control are being evaluated, not just by scientists but by the public, along with the financial costs that can escalate for companies and growers trying to expand their market in the organic area.

Whether mechanical, chemical, cultural, or biological, the goal of weed management should be to reduce or eliminate weeds and limit disturbance as much as possible because weeds most often thrive in disturbed systems.

5 The Need for Change

Crop production is most often conducted on a field scale, and in most cases, inputs are applied at rates averaged for an entire field using equipment that spans multiple crop rows. The needs of individual plants, including weeds, can change dramatically over very short distances. There are obvious requirements of plants, such as nutrients and water, and more subtle requirements, such as light, air, and microbial interactions. In most conditions, plants must compete for resources, which end up diminishing their overall growth and development.

We also know that the strategies that growers use to manage weeds vary between growers, and between and within fields (Riemens et al. 2010). This means standardized or uniform approaches to weed management using emerging technologies are likely to fail in the same way that indiscriminate use of innovative HR products has led to HR weeds. Managing variation in biological systems has to be balanced with managing variation in the social systems or the differences between growers. This may mean targeted communication efforts that address key misperceptions while highlighting the benefits of weed management strategies based on an understanding of the grower situation (Wilson et al. 2009). Increasing the adoption of a dynamic and appropriate management strategy has to be the objective associated with the emergence of new technologies (Hammond et al. 2006).

The potential for new management strategies, a theme of this book, can be found by beginning with an understanding of a commonality of all current weed management strategies. Weeds in production systems often occur in patches of various sizes or as individuals growing among crop plants, yet they are managed in a way that is similar to the crop, large-scale and uniform. A combination of control methods, such as chemical, mechanical, and cultural, is used at different times of the season or over several seasons in most cropping systems, but rarely are single weed plants targeted. Weeds, like crop plants, are not managed at the individual plant scale.

The development of machine-guided technologies for precision weed control has advanced rapidly in recent decades. Technological advancements specific to weed control have been made in many areas, including mechanical, chemical, thermal, and electrical. The first published report of selective spot herbicide application technology was by Lee et al. (1999), who developed a prototype system with microcontroller actuated-specific solenoid valves, delivering liquid to the spray ports, based on the machine vision-generated weed map and robot odometry. Several other weed control tools have been investigated for use in combination with robotic systems, including flame weeding, hot water, organic oils, and high-voltage electrical discharge.

With rapid advances in sensors and guidance technology, potentials for weed control are changing dramatically. By using technologically equipped machinery that can target individual weeds in real time, there is no limit to the number of control tools for use in the field at any one time. The advances in the biological systems engineering field are evidence that "given enough time, an engineer [really] can build anything." Biological research and the latest technological developments in weed control have the potential to radically change the current research approach to weed control and help significantly reduce environmental impacts (e.g., drift, off-target movement) and the high cost of inputs and labor. The potential for developing these precision weed management techniques is real, but challenges remain to do so in a cost-effective manner. Other questions related to scale neutrality or making these innovations available for both small and large operations remain to be addressed.

If it were possible to control weeds without disturbance, the environment would be better off, and growers would have more time to focus on the things that the invention of herbicides allowed for over 50 years ago. It is safe to say that if we could manage weeds without inputting toxins, causing erosion, and changing genetics, we would. Unfortunately, the population of the world continues to increase, yet the amount of arable land available for producing crops will not. Therefore, we need to get more precise in managing crop production and at the same time take steps to protect and limit damage to the ecosystems that ultimately support every single livelihood in every single culture that occupies every single part of the globe.

6 A New Resource

The remainder of this book has been written for the biologist and engineer; the expertise of both is needed to address the current challenges of protecting ecosystems and producing more food for future generations. The discrete and targeted control of weeds in cropping systems using advanced technology is a first step in addressing these challenges.

The six sections of the book include an introduction to the scope of the problem (this chapter) and organic and conventional cropping systems (Chap. 2) (first section). In the second section, a report on the latest advancements in the field of engineering (Chap. 3), a detailed description of weeds and their biology in cropping systems (Chap. 4), and a description of how engineering and weed biology have been combined and the field of biological engineering has advanced (Chap. 5) make up one of the most important sections of the book. In section three, three areas of automated weed control are the focus, including precision planting (Chap. 6), mechanical removal (Chap. 7), and chemical applications (Chap. 8). The fourth section expands the reader's view with examples from the Western Hemisphere (Chap. 9), Western Europe (Chap. 10), and Asia (Chap. 11), of the latest technology that is being used or under development. In the fifth section, the economics of automated weed control (Chap. 12), an industry perspective (Chap. 13), and the potential for automated weed control in underdeveloped countries (Chap. 14) are discussed at length. Finally, the last section (Chap. 15) provides prospects for the future of automation and weed control in precision agriculture.

No other book cuts across two different disciplines with detail and thoroughness to inform readers on the current and provide insight into the future state of weed control. In addition, this book helps to inspire and bring together the next generation of biologists and engineers who are working in the areas of weeds and crop production systems.

References

Beckie H (2006) Herbicide-resistant weeds: management tactics and practices. Weed Technol 20:793–814

Beckie H, Tardif F (2012) Herbicide cross resistance in weeds. Crop Prot 35:15–28

Beckie H, Gill G, Singh H, Batish D, Kohli R (2006) Strategies for managing herbicide-resistant weeds. In: Handbook of sustainable weed management. Food Products Press, New York, pp 581–625

Bhowmik PC (2010) Current status of herbicide resistant weeds around the Globe. J Crop Weed 6:33–43

Donaldson D, Kiely T, Grube A (2002) Pesticides industry sales and usage, 1998 and 1999 market estimates. U.S. Environmental Protection Agency, Office of Prevention, Pesticides, and Toxic Substances, Office of Pesticide Programs, Washington, DC

Gianessi LP, Reigner NP (2007) The value of herbicides in U.S. crop production. Weed Technol 21:559–566

Hammond C, Luschi E, Boerboom C, Nowak P (2006) Adoption of integrated pest management tactics by Wisconsin farmers. Weed Technol 20:756–767

Holm FA, Johnson EN (2009) The history of herbicide use for weed management on the prairies. Prairie Soils Crops 2:1–10, http://www.prairiesoilsandcrops.ca

Lee WS, Slaughter DC, Giles DK (1999) Robotic weed control system for tomatoes. Precis Agric 1:95–113

Lowdermilk WC (1939) Conquest of the land through seven thousand years. U.S. Department of Agriculture's Soil Conservation Service, February 1948 S.C.S. MP-32. First published as USDA Bulletin No. 99. U.S. Department of Agriculture Natural Resources Conservation Service, 1939

Riemens MM, Groeneveld RM, Kropff MJ, Lotz LA, Renes RJ, Sukkel W, van der Weide RY (2010) Linking farmer weed management behavior with weed pressure: more than just technology. Weed Sci 58:490–496

Slaughter DC, Giles DK, Downy D (2008) Autonomous robotic weed control systems: a review. Comp Electron Agric 61:63–78

Timmons FL (1970) A history of weed control in the United States and Canada. Weed Sci 18:294–307

Wilson RS, Hooker N, Tucker M, LeJeune J, Doohan D (2009) Targeting the farmer decision-making process: a pathway to increased adoption of integrated weed management. Crop Prot 28:756–764

Section I
Agricultural Production Systems

Chapter 2
Current State of Weed Management in Organic and Conventional Cropping Systems

Alec F. McErlich and Rick A. Boydston

Abstract Crop losses due to weeds result in reduced yields and quality and increases in harvest costs. Weed management often requires major resource inputs to produce a successful crop. Herbicides are central to the conventional approach to weed management, and they have allowed the grower to reduce management priority, time, effort, and cost of managing weeds. Their use has at times come at a price such as herbicide-resistant weeds, environmental damage, reduced water quality, and loss of genetic diversity. Although growers use a combination of management practices to control weeds, differences between those used in conventional agriculture compared to organic production systems often vary widely in their implementation and relative importance. Approaches to weed management within an organic system revolve around implementing a range of techniques, often consecutively over the course of the cropping rotation. For both organic and conventional growers, weed management remains a significant impediment to optimizing crop yield, improving crop quality, and reducing the costs of production.

1 Introduction

Weeds are ubiquitous to most crops. Most agricultural soils contain millions of weed seed per hectare, and if left unmanaged, weeds greatly reduce crop yields by competing with the crop for nutrients, light, and water. Unlike most other agricultural

A.F. McErlich (✉)
Earthbound Farm, Inc., San Juan Bautista, CA 95045, USA
e-mail: amcerlich@ebfarm.com

R.A. Boydston
US Department of Agriculture, 24106 N. Bunn Road, Prosser, WA 99350, USA
e-mail: rick.boydston@ars.usda.gov

S.L. Young and F.J. Pierce (eds.), *Automation: The Future of Weed Control in Cropping Systems*, DOI 10.1007/978-94-007-7512-1_2,

pests, weeds are present every year in every field and require some degree of management for optimum crop yields and profitability. Weeds comprise the first stage of plant succession following soil disturbance and removal of native vegetation. From the time man first started manipulating crop plants to grow in designated areas rather than gathering food from nature, controlling competing vegetation became a primary task. Planting crops in rows facilitated cultivation and weeding options. Row spacing was largely based on the width of the particular animal or machine that would be used to cultivate the crop.

Crop losses due to weeds vary by crop, weed species, location, and farming system (Bridges 1992; Swinton et al. 1994). Weeds can directly reduce crop yields, reduce crop quality, and increase harvest costs. Weeds not only compete for nutrients, light, and water but can also harbor pests (nematodes, insects, pathogens) of the crop reducing potential yields and quality further (Boydston et al. 2008). Weeds can also reduce the value of the harvested crop such as lowering protein levels in grain and decreasing fruit or seed size. The presence of weeds in the harvested crop may also lower the value of the crop. Jointed goat grass (*Aegilops cylindrica*) in wheat (*Triticum aestivum*) seed, puncture vine (*Tribulus terrestris*) burs and nightshade (*Solanum* sp.) berries in green peas (*Pisum sativum*), nightshade stains on beans (*Phaseolus vulgaris*), and horseweed (*Conyza canadensis*) oil distilled with peppermint (*Mentha piperita*) oil are examples of weeds contaminating and lowering the value of the harvested crop. A Canadian survey of crop losses due to weeds in 58 commodities reported average annual losses of $984 million due to weeds (Swanton et al. 1993). Lentil (*Lens culinaris*) and cranberry (*Oxycoccus* sp.) crops had the greatest percent yield loss due to weeds (25 %), whereas the major crops of corn (*Zea mays*), soybean (*Glycine max*), hay, wheat, potato (*Solanum tuberosum*), canola (*Brassica napus*), and barley (*Hordeum vulgare*) had the greatest monetary value losses.

Most fields are infested with multiple weed species which interact resulting in a combined effect on the crop. Crops vary in their ability to compete and tolerate weeds. Soybean yield was reduced more by weeds than corn yields in previous studies (Swinton et al. 1994). Onions (*Allium cepa*) lack a competitive crop canopy to shade weeds and are susceptible to nearly total crop loss due to uncontrolled weeds (Williams et al. 2007).

2 Changing Consumer Attitude Toward Food

The publication of Rachel Carson's book *Silent Spring* in 1962 is seen by many as the beginning of the modern organic era in the USA. It undoubtedly created a consciousness of environmental issues and food production practices of the time. An increasing consumer awareness of production practices, pesticide residues, food safety, human health, animal welfare, and food quality is largely

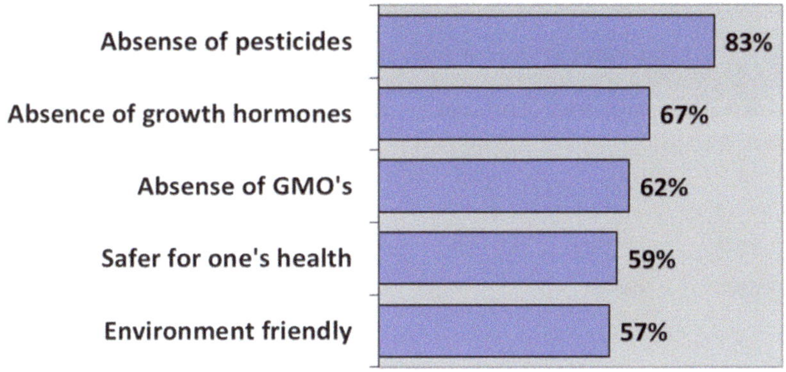

Fig. 2.1 Top five US consumer properties associated with "organic" (Abbrev. from Hartman 2007)

responsible for the increased demand for organic foods (Fig. 2.1). A recent survey of US consumers revealed that nearly two-thirds believe foods are less safe due to chemical use during production and processing (Anon 2010). An increasing number of consumers associate healthy food consumption with improved personal health and wellness. Food is seen as a first step to treating and preventing health problems (Hartman 2010).

Global consumer demand for organic foods continues to increase particularly in Western markets. The higher price of organic foods is often cited as a major reason why consumers avoid buying organic products (Stolz et al. 2011; Sadek and Oktarani 2009). This has led some consumers to seek not only organic foods, but also items labeled as locally sourced, eco-friendly, third party audited, socially responsible, and products produced within sustainable production and processing systems, a trend now termed by many as "beyond organic." Many food and supermarket companies have established production guidelines or require third-party audits of suppliers as part of standard procurement contracts. Third-party verification of production practices is often used to show consumers that a product is produced to set standards by affixing the audit organization logo to the product packaging. Independent third-party verification of production systems is an increasingly common practice within the food industry and has led to an upsurge in the number of various eco-labels worldwide. Examples from the USA are shown in Fig. 2.2. Food and agriculture companies have expanded their corporate social responsibility (CSR) reporting to include not just financial and regulatory information but also measures taken to address sustainable growth, environmental impact, equitable employment policies, and social issues (Martinez 2007).

Organic	Reduced Pesticides	Socially Responsible	Eco Friendly	Sustainable	Animal Welfare
USDA	NutriClean	Fair Trade Certified	RainForest Alliance	Food Alliance	Certified Humane
Canada Organic	Protected Harvest	ProTerra	Salmon Safe	Non GMO Project	Animal Welfare
Demeter	Certified Naturally Grown	Certified Fair Labor	Approved Dolphin Safe	LIVE	American Grassfed
BioGro			Bird Friendly		

Fig. 2.2 Examples of third party food certification labels found in US markets

3 Organic Agriculture

The organic principles involve recognition of the values diversity and natural systems bring to the relationships associated with our use of the planet's resources, crops, and animals to produce food, fiber, and materials. The International Federation of Organic Agricultural Movements (IFOAM) expresses these principles under four core headings of Health, Ecology, Care, and Fairness.

The four principles of organic agriculture as defined by IFOAM (IFOAM 2005) are

- Health—organic agriculture should sustain and enhance the health of soil, plant, human, and planet as one and indivisible.
- Ecology—organic agriculture should be based on living ecological systems and cycles, work with them, emulate them, and help sustain them.
- Fairness—organic agriculture should build on relationships that ensure fairness with regard to the common environment and life opportunities.
- Care—organic agriculture should be managed in a precautionary and responsible manner to protect the health and well-being of current and future generations and the environment.

3.1 Definitions

The International Federation of Organic Agricultural Movements (IFOAM) defines organic agriculture as "a production system that sustains the health of soils, ecosystems and people. It relies on ecological processes, biodiversity and cycles adapted to local conditions, rather than the use of inputs with diverse effects. Organic agriculture combines tradition, innovation, and science to benefit the shared environment and promote fair relationships and good quality of life for all involved."

The early history of organic movement is provided by Conford (2001) and Kristiansen and Merfield (2006). Standards defining the methods by which organic producers operated were first developed in Europe in the 1940s. By the 1970s certification using third-party agencies began to occur, replacing the internal audit systems used by the earliest standards organizations. Today public and private third-party certifiers play a key role to ensure standards are adhered to in all aspects of production through to the customer. Legal definitions and regulations in many counties ensure organic label claims are substantiated by third-party audit programs. A number of countries, including the European Union, USA, Japan, and Canada, have defined "organic" within law and only certified operations may use this term. Seventy-six countries now have some form of organic regulation in statute (Huber 2011). While standards worldwide may differ to some degree, largely in response to local needs and practices, ongoing efforts are underway to establish equivalency agreements in order to harmonize certification and facilitate international trade. As part of the 1990 Farm Bill the US Congress passed the Organic Foods Production Act (OFPA) to enable establishment of the US National Organic Standards. These standards went into effect in April 2001 and are regulated and enforced by the National Organic Program (NOP) of the Agricultural Marketing Service, US Department of Agriculture (USDA).

3.2 World Production

As consumer demand for organic products has increased, so has the land area under organic production. Increasing distribution via mainstream retailers has also driven market growth. Demand from the larger organic market countries of North America, Asia, and Europe has led to an increasing international export trade from African and Latin American countries. Demand for warmer climate crops such as coffee, and the need for counter season supply, has allowed an increasing number of producers in these countries to supply export markets. The rapid consumer demand for organic products in the USA caused periodic product shortages due to supply limitations (Dimitri and Oberholtzer 2009) and as a result led to an increased need for imported products.

The world total area under organic production (agricultural and nonagricultural [beekeeping, wild harvest, forestry, aquaculture, grazed nonagricultural land] production areas combined) reached 80 million hectares in 2010 (Willer and Kilcher 2012). The land area under organic agricultural production increased by 22 % in the 5-year period from 2005 to 2009 and had reached 37.2 million hectares in 2009 (Willer and Kilcher 2011). By 2009, the worldwide number of organic producers totaled 1.8 million, a 400,000 increase from the previous year (IFOAM 2010).

Table 2.1 A world view: regional organic agricultural land area and number of producers 2010 (Willer and Kilcher 2012)

	Agricultural land area (ha, mil)	World organic agricultural area (%)	Number of producers
Africa	1.1	3.0	540,000
Asia	2.8	7.0	500,000
Europe	10.0	27.0	280,000
Latin America	8.4	23.0	270,000
North America	2.7	7.0	17,069[a]
Oceania	12.1	33.0	8,500

[a] 2009

From 2006 to 2010 the global organic food market grew at a compound annual growth rate of 12 %, to reach $59.3 bn. USD in 2010 (Datamonitor 2011). In comparison total world food products were valued at US$4.2 trillion in 2009, representing a 3.7 % compound annual growth rate for the period 2005–2009 (Datamonitor 2011). Sales of organic foods at the retail level have grown by 20–35 % in many countries with a well-developed organic market (Thompson 2000). In 2009 the USA represented the largest organic market ($25.5 bn. USD), followed by Germany ($8.3 bn. USD) and France ($4.3 bn. USD) (IFOAM 2010). Of the total world food market, organic sales are less than 3 % (Thompson 2000) (Table 2.1).

3.3 *World Land Use and Crops*

Of the 37.2 million hectares under organic management in 2009, 47 % was classified as agricultural lands, i.e., cropland, permanent grassland, and other agricultural land. The remaining 53 % was utilized for forestry, beekeeping, aquaculture, and grazing on nonagricultural land (Willer and Kilcher 2011). In 2009, 0.9 % of the world's agricultural land was managed using organic practices.

The major arable crops in organic agriculture are cereals (including rice), green fodder, oilseeds, vegetables, and protein crops. The area in cereals is underreported as producer countries such as India, China, and the Russian Federation did not report data (Willer and Kilcher 2011). The major permanent crops are coffee, olives, cocoa, nuts, and grapes (Willer and Kilcher 2011). The total arable land in organic production accounts for 0.4 % of the world's total arable land area.

3.4 *Weed Management*

Although some see organic agriculture as returning to the techniques of an earlier era, it is in practice a combination of traditional methods along with modern innovation and science. Organic agriculture is a set of management practices that rely on integrating natural cycles, enhanced biodiversity, preservation of natural resources, and improved soil health and quality, in order to promote ecological balance. It is a system that provides a range of ecosystem services such as improved soil tilth, lower energy use (Reganold et al. 2001; Refsgaard et al. 1998; Gündoğmus 2006; Mäder et al. 2002), reduced pesticides (Mäder et al. 2002), greater biodiversity (Mäder et al. 2002; Letourneau and Bothwell 2008), and increased carbon sequestration (Pimentel et al. 2005).

Weed control methods are broadly categorized as mechanical, biological, cultural, chemical, and preventative. Although most farmers may use a combination of management practices to control weeds, the approaches and priority ranking placed against each practice used in conventional and organic agriculture are often very different. Weed management in organic systems revolves around implementing a range of techniques within an integrated natural-based system, rather than reliance on a single or narrow selection of management techniques, typical of industrial agriculture. Some describe this as the "many hammers" approach as several actions are taken, often consecutively, over the course of the cropping rotation to reduce the weed burden. Within conventional agriculture the overwhelming reliance is on the use of herbicides to largely eliminate weed competition. While management practices, such as the use of crop rotation and cover crops, are widely used by mainstream agriculture, their primary functions are not directed toward weed management. Organic growers on the other hand recognize these practices as a core function of a good weed management program.

Of all the production challenges facing organic growers, weed management remains for most one of the most difficult, frustrating, expensive, and time-consuming management aspects of producing a crop for market. Despite an increasing selection of cultivation equipment and an improved understanding of weed management techniques and weed ecology, growers continue to find weed management a significant impediment to optimizing yield, quality, and income. Grower concerns over an inability to control weeds are often cited as the leading deterrent to converting to organic production. Numerous grower surveys of organic agriculture demonstrate that weed control remains a major and enduring challenge (Walz 1999; Kristiansen et al. 2001; Birzer and Badgery 2006; Anon 2007). This is not to say that all organic growers struggle with controlling weeds. Those growers who have implemented successful management techniques have done so through hard work, a sharp learning curve, innovation, and a firm decision not to allow weeds to develop mature seed. With time, skill, and effort, growers are able to reduce the weed seed bank to very low levels, which has resulted in significant financial benefit.

For many organic growers weed management remains one of the most resource-intensive management activities from the perspective of time, effort, input costs, potential impact on crop yield and quality, capital investment, and energy consumption. There are few times of the year when a grower is not actively working on some aspect of weed management.

Although only the US Department of Agriculture, National Organic Program (USDA-NOP) standards relating to weed management are shown, most organic

Table 2.2 USDA-NOP standard section addressing weed management from §205.206 crop pest, weed, and disease management practice standard

Category	Options
Management practice	Crop rotation
	Soil and crop nutrient management
	Cultural choices (e.g., resistant to prevalent pests)
Pest problem (weeds)	Mulching with biodegradable materials
	Mowing
	Livestock grazing
	Hand weeding and mechanical cultivation
	Flame, heat, or electrical means
	Plastic or other synthetic mulch (removed at end of season)
	Biological or botanical substance (after other options)

standards in use around the world employ a similar code of practice in relation to weed management. The relevant section of the USDA-NOP standards relating to weeds (see Table 2.2) directs producers to use cultural and physical tools before utilizing any approved pesticides.

3.5 Herbicides Accepted for Weed Management in Organic Agriculture

An increasing number of herbicides are permitted for use in organic agriculture. These materials are based on naturally occurring compounds such as plant oils, corn gluten meal, fatty acids, acetic acid, and biological materials. Most products are categorized as plant oil extracts and act as a nonselective contact on green vegetation. Corn gluten meal acts to limit the normal development of plant roots. An acetic-acid-based product is approved for use in the USA. Soaps are used to control moss and algae. Because of the high cost of these materials, their use is limited to directed or spot sprays in higher-value crops. High spray volumes must be used to ensure adequate coverage of the target weeds. Lanini (2010) describes some characteristics of plant oil-based herbicides:

- Generally regarded as minimum risk pesticides
- Act only on small, newly emerged weeds
- Have greater efficacy against broadleaf weeds than grasses
- Require good coverage and repeat applications to achieve adequate control
- Require higher spray volumes at lower concentrations to achieve adequate control
- Lack residual action
- Require an approved adjuvant to improve efficacy
- Provide a greater level of weed control when applications occur at warm temperatures

4 Conventional Cropping Systems

Throughout history, the approach to agricultural weed management has largely been influenced by the energy source for those activities. The progression from human to animal to mechanical to chemical has been influenced by the technologies of the period. In modern agricultural systems, the energy source is largely obtained from fossil fuels. Modern, industrialized approaches to agricultural production rely on high input systems, centered on the use of fossil-fuel-based inputs such as fertilizer and pesticides, to produce high yields. Improvements in plant breeding and irrigation practices have also contributed to dramatic yield increases over several decades. There is an increasing realization that this approach has come at a price. Overuse and inappropriate use of fertilizers and agrochemicals are associated with loss of genetic diversity, reduced soil and water quality, and are linked with human health problems. Central to weed management within conventional agriculture is the use of herbicides. Herbicides often allow the producer to reduce the management priority, time, effort, and cost of managing weeds.

4.1 Weed Management

In conventional cropping systems, weeds are viewed as a pest that must be managed if maximum yields and profits are to be realized. Due to the high level of weed control obtained with synthetic herbicides, grower's expectations and goals of weed-free fields have become the norm, sometimes without warrant with regard to the level of weed control required to eliminate negative effects on crop yield. As in organic systems, weed management in conventional cropping systems relies on a combination of cultural, mechanical, chemical, and biological methods, but because of the great degree of efficacy and selectivity of modern synthetic herbicides and the simplicity of their use, conventional agriculture relies heavily on chemical weed control methods. Prior to the 1950s most weed control in cropping systems was accomplished with a combination of mechanical cultivation tools, hand weeding, and cultural practices. Weed control was a major task for producers requiring a major portion of their time in the field. As synthetic herbicides became available, growers discovered they could control most weeds in a fraction of the time and with a much higher level of control than with cultivation. In- row weed control, often difficult and expensive with mechanical means, became easily accomplished with synthetic selective herbicides. Crafts (1960) and Timmons (1970) provided detailed histories of the twentieth-century weed control developments in the USA and Canada. Following the discovery of 2,4-D in the early 1940s, the number of herbicides used or tested in the USA and Canada increased from 15 to 120 in 1969 (Timmons 1970).

Modern conventional agriculture relies heavily on synthetic herbicides to manage weeds due to the relatively low cost, ease of use, and high efficacy.

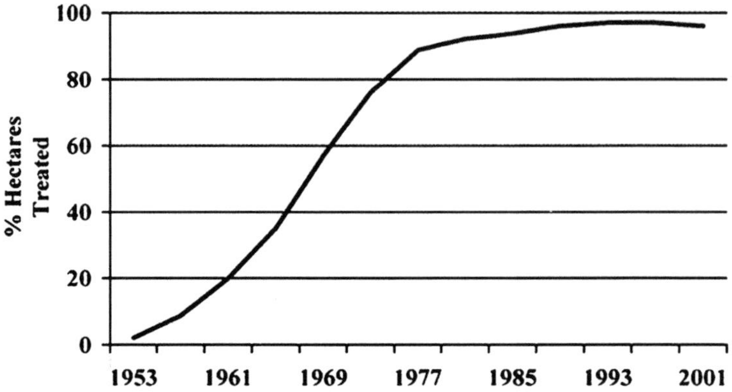

Fig. 2.3 Land area treated with herbicides in six major U.S. crops (Gianessi and Reigner 2007) (Note: Trendline of corn, cotton, soybean, sugarbeet, peanut, and rice. Source: Brodell et al. 1955; Strickler and Hindson 1962; USDA 1968, 2001; Eichers 1978; Duffy 1983; Andrilenas 1975)

Herbicides account for approximately two-thirds of expenditures for all pesticides used in US agriculture (Gianessi and Reigner 2007). The herbicide 2,4-dichlorophenoxyacetic acid (2,4-D) was the first widely used synthetic herbicide in modern agriculture (Timmons 1970). It was first commercialized in the late 1940s and was rapidly adapted by producers of corn, wheat, rice (*Oryza sativa*), and other grass crops for broad-spectrum broadleaf weed control. 2,4-D was relatively inexpensive, easy to apply, and controlled many problem broadleaf weeds that were often difficult to manage with cultivation and other methods. Perennial weeds, such as Canada thistle and field bindweed, were particularly difficult to control before the advent of 2,4-D, requiring repeated cultivation and often sacrificing an entire growing season without a crop to reduce the weed pressure with repeated tillage (Freed 1980).

The 1950s saw the introduction of the triazine family of herbicides, including atrazine and simazine. Atrazine and simazine were widely used in crops such as corn and fruit trees for preemergence weed control. These herbicides were persistent in the soil and provided control of most annual weeds for several months. Despite the weed control benefits of atrazine, its relatively high mobility in the soil coupled with its long persistence and extensive use led to atrazine contamination of groundwater in many production areas. In addition, repeated use of atrazine eventually led to selection of herbicide-resistant weed biotypes (Bandeen et al. 1979; Heap 2011).

Increases in herbicide use led to increases in pesticide use in agriculture for many years. As a percentage of total pesticides applied, herbicides rose from 33 % in 1966 to 70 % in 1986, and herbicides still dominate in terms of kilograms of active ingredients applied. Land area treated with herbicides in six major crops tripled from the early 1960s to the early 1970s as farm size increased and pesticide companies developed many new selective herbicides Fig. 2.3. Over 90 % of (the six major crops corn, cotton, soybean, sugar beet (*Beta vulgaris*), peanut (*Arachis hypogaea*) and rice) are treated with herbicides (Gianessi and Reigner 2007),

Table 2.3 World and US pesticide amount by pesticide type—2007 (Grube et al. 2011)

	World market		US market	
Type	Mil kg a.i.	%	Mil kg a.i.	%
Herbicides[a]	951	40	241	47
Insecticides	405	17	42	8
Fungicides	235	10	32	6
Others[b]	773	33	199	39
Total	2,363	100	514	100

[a] Herbicides include herbicides and plant growth regulators

[b] Others include nematicides, fumigants, and other miscellaneous conventional pesticides, and other chemicals used as pesticides such as sulfur, petroleum oil, and sulfuric acid

and corn and soybean production accounts for the majority of US herbicide use (Vecchia et al. 2009).

Glyphosate was introduced in 1974 for nonselective weed control. Glyphosate allowed growers to control weeds without preplant tillage and was crucial for the adoption of reduced conservation tillage systems. Most growers of the major row crops have adopted reduced or no-till production practices that reduce soil erosion and help maintain soil organic matter and rely on contact or burn down type herbicides, such as paraquat and glyphosate. Glyphosate is now the major herbicide used in selective weed control due to the development of crop plants that are resistant to it. Although less soil erosion occurs in no-till or reduced till systems, increased use of herbicides is also common, which can lead to selection for herbicide-resistant weeds.

The amount of pesticide used in the USA accounted for 23 % of total world pesticide amount used and 25 % of world herbicide amount used (Grube et al. 2011). Herbicides still account for the largest portion of the USA and world pesticide use (Table 2.3). Herbicides used in the USA in 2007 exceeded 240 million kg and accounted for 47 % of the total pesticide use (Table 2.3). In part due to the widespread use of herbicides and excellent efficacy, herbicide research has dominated the discipline of Weed Science for several decades (Zimdahl 2000).

Herbicides have brought numerous benefits to farmers including high efficacy, reduced trips through the field, ease of use, lower labor and cultivation costs, less time spent on weed control, and reduced soil erosion. Despite the numerous benefits of herbicides, there are undesirable effects associated with herbicide use including environmental and health issues. Herbicide-resistant weeds, misapplication, off-target movement, and persistence of some herbicides in soil can injure susceptible crops. Environmental concerns including herbicide contamination of ground and surface waters have led to attempts to reduce herbicide inputs in agriculture. Alternative weed management practices (cultivation, crop rotation, cover crops, mowing, mulching, etc.) are typically used in combination with herbicides in most production systems. Banding of herbicides in the crop row combined with cultivation between the rows rather than broadcast spraying is sometimes used to reduce herbicide inputs.

New sprayer technology is available that only applies herbicide when a weed is identified, but has had limited adoption by growers to date. One example of this technology is the WeedSeeker® sprayer which utilizes optics and computer circuitry to detect green plants and activate spray nozzles to apply a nonselective herbicide. Sprayers equipped with this technology are able to reduce the total amount of herbicide applied when weed canopies do not entirely cover the ground and have been utilized primarily for controlling weeds in fallow ground.

The heavy reliance of conventional agriculture on synthetic herbicides has decreased grower knowledge and use of alternative weed control practices (mechanical and cultural) that are standards in organic production. Quite often when new weed problems emerge, the mind-set of conventional grower is to look for the next best herbicide treatment rather than to evaluate the possible causes of the problem and adjust cultural or other weed management practices accordingly.

New weed management options are needed to prevent and delay development of herbicide-resistant weeds. Simply rotating herbicide mode of action is often effective, but not in all cases, and multiple resistance to herbicides with different modes of action has developed in numerous weed species. Growers must adopt and develop other technologies for weed control that compliment and extend the utility of the herbicide tools they now have. Continuous, repeated, and exclusive selection pressure on weeds from any weed control tactic will select for weeds resistant or tolerant to the specific control tactic. Weed community shifts can occur as a result of natural tolerance to the herbicide used or through an avoidance strategy related to weed emergence timing. An integrated approach to weed management is required to prevent or retard the development of resistant weed populations. Use of diverse crop rotations and cultural practices that suppress weeds and mechanical, biological, and chemical control methods all contribute to slowing or preventing the development of resistant weeds. New technologies such as robotic weeders could be used to compliment current weed control methods by eliminating seed production of escape weeds and providing new selection pressure unrelated to that of herbicides.

4.2 Herbicide-Resistant Crops

Major acreage crops such as field corn, soybeans, wheat, and cotton drive most of the development of new herbicides as the high development, registration, and regulatory costs can be more readily recuperated with the large sales volume. More recently, seed companies have developed herbicide-resistant traits for many of the major acreage crops. Grower adoption of genetically modified crops with resistance to glyphosate has occurred rapidly in corn, soybeans, cotton, canola, sugar beets, and alfalfa (*Medicago sativa*). The broad spectrum of weed control, low cost, ease of use, and reduction in cultivation when using this technology has resulted in the most rapidly adopted technology in the history of agriculture (Dill et al. 2008).

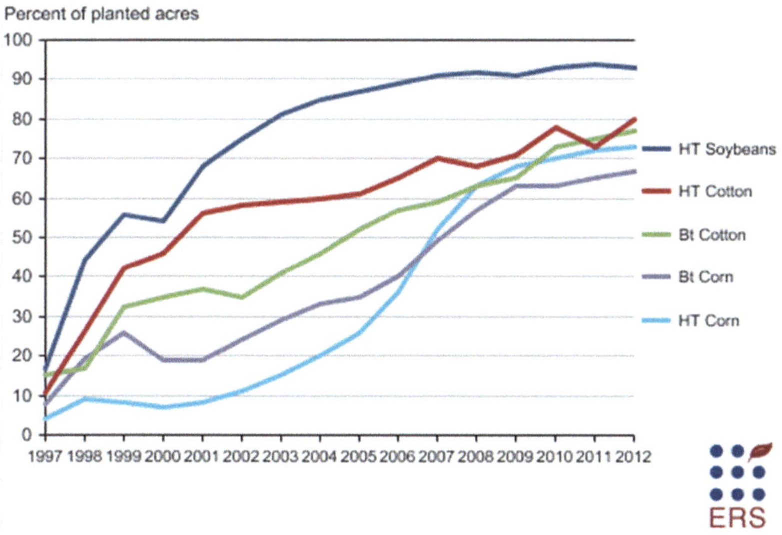

Fig. 2.4 Adoption of genetically engineered crops in the US (USDA ERS 2011) (Data for each crop category include varieties with both HT and Bt (stacked) traits. Source: USDA, Economic Research Service)

In some cases, the herbicides used in herbicide-resistant crops are more convenient to use, less toxic to mammals, and less persistent in the environment (Cerdeira and Duke 2006). Initially, glyphosate was extremely effective in controlling weeds in glyphosate-resistant crops, and many growers relied exclusively on glyphosate to manage weeds. Due to the intense selection pressure placed on weed populations with the widespread use of glyphosate, weed resistance to glyphosate has become more prevalent. Weed management programs cannot rely on single tactics, whether an herbicide or other form of weed control, or weeds will ultimately adapt and survive in large numbers. Weed resistance to glyphosate has forced many growers to rotate to or tank mix with herbicides having a different mode of action. Combining herbicides, cultural practices, and mechanical tactics provides the greatest protection from herbicide-resistant weeds and is part of an integrated weed management program.

Most high-value fruit and vegetable crops are grown on fewer hectares and have not had the luxury of new herbicide development due to the lower market potential and higher risk of crop injury. As a result, herbicide development for these crops usually consists of adapting and registering herbicides that were developed for one of the major field crops. This often results in fewer and older herbicides labeled on these higher-value crops and sometimes no herbicides available for control of certain problem weeds. Consumers have also voiced concern about the safety of genetically modified crops to humans and the environment. This, coupled with the lower profit incentive due to lower acreage, few seed companies have developed herbicide-resistant fruit and vegetable crops. As a result, many specialty fruit and vegetable growers rely more heavily on tillage, cultivation, mulching, and hand weeding to manage weeds (Fig. 2.4).

Risks associated with herbicide-resistant crops include marketing problems with grain contamination, segregation and introgression of herbicide-resistant traits, marketplace acceptance, and an increased reliance on herbicides for weed control. The evolution of glyphosate-resistant Palmer amaranth (*Amaranthus palmeri*), common water hemp (*Amaranthus tuberculatus*), creeping bent grass (*Agrostis stolonifera*), horseweed (*Conyza canadensis*), rigid ryegrass (*Lolium rigidum*), and goose grass (*Eleusine indica*) is directly attributable to the adoption of glyphosate-resistant crops and the concomitant use of glyphosate as the primary herbicide for weed management (Green and Owen 2011; Owen and Zelaya 2005).

5 Weed Management Challenges in Organic and Conventional Systems

The rate of conversion to organic production in the USA has not kept pace with consumer demand, despite significant price premiums for many organic crops. Grower concerns about converting to organic agriculture are in part associated with the steep learning curve needed to acquire new knowledge and expertise, lower potential yields during the initial years, a lack of an organic price premium during the transition years, and the need to acquire an understanding of the functionality of the markets and companies involved in moving crops into the consumer market. Growers are concerned for the economic risks associated with converting their farm to an organic system (McCann et al. 1997). It may take several years before crop performance reaches parity with pre-conversion yields.

While conventional growers place considerable emphasis on herbicides to manage weeds, both conventional and organic growers recognize cultivation as an important management strategy (Doohan et al. 2010). A high proportion of organic farmers often associate weed problems on their farms with a failure of good weed management practices on adjacent land while downgrading their own poor practices as the cause of the weed problem (Doohan et al. 2010). A lack of good weed identification and taxonomy skills by growers often hinders good weed management decisions. The consistent and timely implementation of a diverse range of management practices allows the organic grower to minimize the impact of weeds within a rotation and over time reduce weed control costs to a minor portion of the total costs of production (Fig. 2.5).

The availability of herbicides often provides the conventional grower an opportunity to reduce the degree of seedbed preparedness compared to that required for optimal establishment of organic crops. Consistency of soil tilth and surface evenness are core requirements to optimizing the performance of inter-row weeding implements after seeding. Growers often require time to learn the finer aspects of seedbed preparation in order for later weed cultivations to be performed with a high degree of accuracy and efficiency. Often excess soil volume from the between rows, deep channeling caused by seed drill press wheels, and/or clods will hinder the desired flow of soil off cultivator tines or knives. This often results in the crop being

Fig. 2.5 Examples of weed management practices utilized by organic growers

submerged by excess soil and results in plant loss or delayed plant development. Here the crop is most susceptible to loss at the earliest growth stage and the initial cultivation following seeding.

5.1 Hand Weeding

The term hand weeding applies to not just physically pulling weeds by hand but also the use of hand tools to manage weeds. Artifacts show hand hoes have been used for thousands of years in the cultivation of crops important to humanity. The sixth Babylonian King Hammurabi mentions them in the Code of Hammurabi around 1700 BCE. Figurines from the Tang Dynasty 600–900 CE depict farmers holding hand hoes. Today the hand hoe continues to play an important role where row crops are grown, be they in traditional cultures or industrial agriculture.

Hand weeding often remains a core means to control weeds particularly for small area organic growers, where perhaps high capital costs and lack of size-appropriate equipment negate investment in specialized cultivation equipment. This is not to say large area or high-value crop growers do not use hand labor. In large-scale systems there is a need to invest in weeding equipment in order to cover large areas, within short periods of time, when optimal weed control opportunities exist. For conventional onions, carrots, mint, and grass seed growers, hand weeding is still an important option to controlling weed escapes and volunteer crops (Williams and Boydston 2005). Growers will use hand labor to remove potential contaminant weeds prior to harvest or to remove immature seeds before seed is added to the soil seed bank. A typical example involves processing green peas (*Pisum sativum*) where Canada thistle (*Cirsium arvense*) flower buds are potential contaminants at harvest. In addition, hand removal of weeds that escape common herbicide programs in conventional crops is an important and effective strategy to prevent development of herbicide-resistant weed populations.

The decision to use hand labor is typically dictated by the need for botanical purity in the resulting crop. Growers producing intensive crops with stringent quality requirements will often utilize hand labor to remove potential contaminant weeds prior to harvest. This is particularly important in herb and fresh salad crops, for example, where the need for botanical purity is a critical requirement leading up to harvest and where the tolerance for extraneous vegetative matter within the crop is extremely low.

Access to a skilled labor force is an ongoing concern for most growers as costs increase and immigration requirements are intensified on foreign labor. With the need to remove weeds at an early growth stage, often at a time when the crop is also just beginning to develop, a trained hand crew can make or break a crop. Distinguishing between weed and crop at an early growth stage is frequently a challenge, and if mistakes are made they are often costly. Combined with the cost of the crew, hand weeding can often represent the most costly input to a crop.

Hand weeding is time consuming, slow, and physically demanding. In 1975 the California Supreme Court (Murray 1982) ruled to ban the use of the short-handled hoe (el cortito, the short one) due to lower back injuries caused by the stooped working position workers endured. To ease the physical burden, mobile (e.g., Drängen) or tractor-mounted multi-row platforms on which workers lie or sit, are used by some row crop growers. New technologies that can kill or suppress weeds with the precision of human labor while reducing the associated costs and requirements are important for the sustainability of organic and conventional production systems.

5.2 Crop Rotation

The ability to rotate crops from one field to another over several years is a corner stone of annual cropping systems in organic and many conventional cropping systems. In a survey of growers, organic producers are more likely to mention rotation as a useful weed management tool (Doohan et al. 2010). More complex (diversified) rotations build multiple stress points against weeds. By growing crops with different management requirements, it is difficult for weeds to become a dominant factor within the rotation. The use of weed-suppressive crops can suppress weed growth following grain harvest by outcompeting the weeds for light, nutrient, and water resources. An example is the undersowing of cereals with red clover (*Trifolium pratense*). Increased crop and spatial diversity within a rotation allows the grower to alter the timing and types of agronomic operations throughout the growing season.

Increasing worldwide consumer demand for organic products has led to a commensurate increase in the land area under organic management and the number of organic farmers. A large portion of the growers are either new to farming, new to organic production, or both. The learning curve for these growers is often steep, not only involving production practices but also understanding access to the organic market and the companies involved in taking organic products into the marketplace. The challenge for growers is to implement a crop rotation based on agronomic and management requirements while also taking into consideration the marketability of the selected crops. Ability to access markets is often dictated by growing region and size of the farming and/or processing and/or the consumer base.

5.3 Plant Breeding and Variety Selection

Over thousands of years nature has demonstrated that adaptability is crucial to survival. Organic growers recognize the importance of diversity within agricultural systems, and this extends to the crop varieties grown. Critics often claim that yields from organic systems are too low to adequately feed a growing human and animal population compared to those obtained from conventional agriculture (Smil 2001). The majority of varieties used in organic agriculture come from conventional breeding programs, and their use is often based on traits considered valuable in an organic system. Only recently have we seen the development of breeding programs within and for organic agriculture. Murphy et al. (2007) demonstrated how soft white winter wheat (*Triticum aestivum*) varieties do not perform equally across organic and conventional agricultural systems. Few plant breeders have focused breeding efforts on weed-suppressive qualities or crop tolerance to weeds. Several factors associated with crop canopy development near emergence, canopy closure, and the reproductive period influence the ability of sweet corn (*Zea mays*) to suppress and compete with wild millet (*Panicum miliaceum*) (Yim et al. 2009). Murphy et al. (2008) compared historical and modern spring wheat varieties, for tolerance to mechanical cultivation, and found significant differences in weed biomass.

Breeding for faster emergence, improved seedling vigor, and tolerance to seedling establishment diseases in annual crops would ensure a more consistent plant stand that meets all of the other crop requirements. Crop planting date decisions are not always made by the grower, particularly in process vegetable production where the processing company typically dictates the crop sowing date. Early season crops such as green peas (*Pisum sativum*) are often sown into cool soils in the Pacific

Northwest region of the USA, a major pea growing area, in an effort to avoid contaminant weeds such as nightshade (*Solanum* sp.) in later sown crops. Few organic seed treatments are effective at low temperatures soil. Growers typically increase seeding rates by 20–40 % to compensate for expected plant loss due to disease and post-planting weed cultivation. Poor crop plant populations create an opportunity for weeds to fill the vacant space.

Growers often find the decision to initiate the first mechanical weeding operation a difficult one when crop emergence and early development occur over several days. With individual plants at different growth stages, growers frequently wait for the later emerging plants to reach a stage tolerant of mechanical weeding. Because these plants tend to be less vigorous, their initial growth rate is slow and more susceptible to damage or loss if cultivation occurs too early. Meantime weed development continues and often progresses to a growth stage beyond that for effective removal by weeding implements.

5.4 Mechanical Cultivation

Primary tillage is used to meet a range of objectives, including managing plant residues, weed control, warming cold soils, improving water penetration, alleviating hard pans, seedbed preparation, incorporation of fertilizer and compost, and aerating the soil. Without due care tillage can lead to soil erosion and a reduction of soil quality. Conservation practices to counter the negative factors of tillage enable soil qualities to be maintained or improved over time. Practices such as incorporating cover crops/residue, addition of compost and manures, contour tillage, and implementing sound crop rotations all have a role in conserving soils.

Limitations to mechanical cultivation include dealing with excess crop residue, weather related soil conditions, limited in row weed control, soil type, variations in crop growth, degradation of soil quality and health, operator skill, and often redistribution of weed seeds within the soil profile. Cultivation implements often require significant capital investment by the grower.

Organic growers rely on tillage as the principal method of weed control in row crop production. The types and methodology of implements used for weed management are detailed in two books: *Steel in the Field*, Bowman (1997), and *Practical Weed Control*, van der Schans and Bleeker (2006). When discussing equipment many different tactics and viewpoints are held by growers, each adapting from personal experience and situation. The introduction of organically acceptable herbicides and the recent development of roller-crimper equipment to destroy living cover crops in situ are allowing organic growers to consider the advantages of minimized tillage, strip-till and no-till systems. Utilizing precision strip tillage equipment over controlled traffic lanes may offer improved weed control in minimum tillage situations (Kurstjens 2007).

Following planting, strategies to control weeds are generally limited to mechanical and cultural methods. Between-row weed control is a relatively easy procedure in annual row crops grown at wide spacings, using undercutting sweeps and shovels. Cultural methods such as fast canopy closure, use of buried drip irrigation tape, and directed fertilizer applications can all assist to reduce weed growth between crop rows. The intra-row (within row) weeds are the more difficult to control particularly in crops grown at high density. Intra-row weeds also tend to reduce yields the most due to their proximity to the crop. Most hand weeding used in production fields is aimed toward these intra-row weeds. Regardless of the crop, weeds situated within the crop row are the most challenging to control particularly as crop density increases for vegetable crops such as carrots and onions. Controlling intra-row weeds and reducing hand weeding costs are primary objectives that new technologies can address for organic and conventional production practices.

Although removal of weeds is nearly always the ultimate goal, this may not always be possible when considering intra-row weeds. In some cases, the aim may be to give a competitive advantage to the crop by using transplants, preemergent flaming of weeds, plastic mulches, and stale seedbed techniques. Delaying weed development by smothering with cultivated soil may be an option in large seeded crops such as sweet corn (*Zea mays*) and beans (*Phaseolus vulgaris*). Establishing a dry soil surface mulch by careful irrigation management can reduce the number of weeds germinating from the first few centimeters of soil. During cultivation the dry soil flows off the ends of cultivator knives or tines in a free-flowing wave that, given correct forward tractor speed and equipment modification, allows the soil to move into the crop row and smoother small weeds.

Depleting the weed seed bank near the soil surface using a preplant weed flush can reduce the potential number of weeds emerging during crop establishment. Shallow cultivation or flame weeding equipment is used to minimize the movement of deeper profile, weed seed laden soils into the crop germination zone. Most growers also use this same strategy when sufficient time allows for weed germination to occur before sowing the following crop or during fallow periods.

Tillage and mowing remain the most widely utilized management tools for perennial weeds. Most techniques rely on reducing the ability of plants to produce or store food, either by removing vegetative growth or tilling the soil to expose roots and rhizomes to desiccation or freezing. In organic agriculture, perennial weed control typically involves multiyear programs employing tillage, fallow periods, competitive crops, and vegetation control and minimizing movement of reproductive material around the farm. Even with herbicides, perennial weeds often represent the most serious and difficult to control weed category for conventional and organic growers.

6 Conclusion

Despite a more than doubling of organic farmland in the USA between 1997 and 2005, consumer demand for organic products has at times exceeded supply for some products. Grower concern over weed management remains a major hurdle for existing organic growers and hinders the conversion rate of conventional land to organic production. Organic farmers face the same complex issues as conventional famers when confronted with weed management decisions, but to obtain or maintain certified organic status a more holistic approach is required. From a weed management perspective, farmers who demonstrate proficiency, are timely, understand weed ecology and biology, develop a contingency plan for every field, and have a longer-term vision for the farm are better able to manage weeds. Additional focus on improved efficiencies from farm to consumer, reduced costs of production, improved technology transfer, and enhanced yield and quality are necessary if organic producers are to meet continuing consumer demand for organic foods.

The increasing problem of herbicide-resistant weeds has contributed to a reassessment of alternative weed control methods including cultivation to ensure adequate control. Conventional growers often acquire organic production knowledge in order to overcome some of the limitations of an intensive herbicide system. Where farmers operate both conventional and organic systems, many recognize that new skills obtained from organic agriculture have application and benefits across their entire farming operation. A slowing of the rate, at which new herbicides are commercialized, input costs, and the limitations posed by a lack of label registrations on minor crops are encouraging growers to look beyond synthetic chemistry for answers to help with weed management. Uncertainty over access to adequate sources of labor to undertake weeding by hand and increasing labor rates are major issues facing growers and will necessitate new technology. Multiple weed control methods used at multiple spatiotemporal scales are necessary if we are to reduce the negative impacts of weeds on crop production. Utilizing an integrated approach by combining a greater number of existing management options will be necessary if we are to preserve agroecosystems and supply adequate food to a growing world population. New technologies offer promise to meet many of the challenges in weed control, both in organic and conventional production systems.

References

Anon (2007) Experiences and outlook of Minnesota organic farmers. Minnesota Department of Agriculture. http://www.mda.state.mn.us/news/publications/food/organicgrowing/2007orgsurvresults.pdf. Accessed 8 Nov 2012

Anon (2010) Health and wellness trends report 2010. Natural Marketing Institute, Harleysville

Bandeen et al (1979) Annual meeting of the Weed Science Society of America 1979 (Abstract), p 108

Birzer T, Badgery W (2006) Organic wheat farming in Australia: issues farmers face. AFBM 3(2):1–13

Bowman G (ed) (1997) Steel in the field; a farmer's guide to weed management tools, Handbook series book 2. Sustainable Agriculture Network, National Agricultural Library, Beltsville

Boydston RA, Mojtahedi H, Crosslin JM, Brown CR, Anderson T (2008) Effect of hairy nightshade (*Solanum sarrachoides*) presence on potato nematodes, diseases, and insect pests. Weed Sci 56:151–154

Bridges DC (1992) Crop losses due to weeds in the United States – 1992. Weed Science Society of America, Champaign

Carson RL (1962) Silent spring. Houghton Mifflin, Boston
Cerdeira AL, Duke SO (2006) The current status and environmental impacts of glyphosate-resistant crops: a review. J Environ Qual 35:1633–1658, Reviews and analyses
Conford P (2001) The origins of the organic movement. Floris Books, Edinburgh
Crafts AS (1960) Weed control research: past, present and future. Weeds 8:535–540
Datamonitor (2011) Global: food, beverages and tobacco industry guide
Dill GM, CaJacob CA, Padgette SR (2008) Glyphosate-resistant crops: adoption, use and future considerations. Pest Manage Sci 64:326–331
Dimitri C, Oberholtzer L (2009) Marketing US organic foods: recent trends from farms to consumers, Economic information bulletin number 58. Economic Research Service, United States Department of Agriculture, Washington, DC
Doohan D, Wilson R, Canales E, Parker J (2010) Investigating the human dimension of weed management: new tools of the trade. Weed Sci 58:503–510
Freed VH (1980) Weed science: the emergence of a vital technology. Weed Sci 28:621–625
Gianessi LP, Reigner NP (2007) The value of herbicides in U.S. crop production. Weed Technol 21:559–566
Green JM, Owen MDK (2011) Herbicide-resistant crops: utilities and limitations for herbicide-resistant weed management. J Agric Food Chem 59:5819–5829
Grube A, Donaldson D, Kiely T, Wu L (2011) Pesticide industry sales and usage: 2006 and 2007 market estimates. EPA's Biological and Economic Analysis Division, Office of Pesticide Programs, Office of Chemical Safety and Pollution Prevention, U.S. EPA. http://www.epa.gov/opp00001/pestsales/07pestsales/market_estimates2007.pdf. Accessed 14 July 2011
Gündoğmus E (2006) Energy use on organic farming: a comparative analysis on organic versus conventional apricot production on small holdings in Turkey. Energy Conserv Manage 47:3351–3359
Hartman (2007) The organic marketplace. The Hartman Group, Bellevue
Hartman Group (2010) Reimaging health and wellness: lifestyle and trends report 2010. The Hartman Group Inc. , Bellevue
Heap I (2011) The international survey of herbicide resistant weeds. http://www.weedscience.com. Accessed 14 July 2011
Huber B (2011) The world of organic agriculture: regulations and certification emerging trends 2011. Paper presented at BioFach Congress, Nürnberg, 16–19 February 2011
IFOAM (2005) International Federation of Organic Agricultural Movements, Annual reports 2005–2006, Bonn, Germany
IFOAM (2010) International federation of organic agricultural movements, annual report 2010, Bonn
Kristiansen P, Merfield C (2006) Overview of organic agriculture. In: Kristiansen P, Taji A, Reganold J (eds) Organic agriculture, a global perspective. CSIRO Publishing, Collingwood, pp 1–19
Kristiansen PE, Sindel BM, Jessop RS (2001) Weed control in organic horticultural cropping systems. In: Organic farming workshop, NSW Agriculture Centre for Organic Farming, Bathurst, 5–7 July 2001
Kurstjens DAG (2007) Precise tillage systems for enhanced non-chemical weed management. Soil Tillage Res 97:293–305
Lanini WT (2010) Organic herbicides – do they work? Calif Weed Sci Soc J 6(1):1–3
Letourneau DK, Bothwell SG (2008) Comparison of organic and conventional farms: challenging ecologists to make biodiversity functional. Front Ecol Environ 6:430–438
Mäder P, Fliebach A, Dubois D, Gunst L, Fried P, Niggli U (2002) Soil fertility and biodiversity in organic farming. Science 296:1694–1697
Martinez SW (2007) The U.S. food marketing system: recent developments, 1997–2006, Report number 42. United States Department of Agriculture, Economic Research Service, Washington, DC
McCann E, Sullivan S, Erickson D, De Young R (1997) Environmental awareness, economic orientation, and farming practices: a comparison of organic and conventional farmers. Environ Manage 21:747–758

Murphy K, Campbell KG, Lyon S, Jones SS (2007) Evidence for varietal adaptation to organic farming systems. Field Crop Res 102:172–177

Murphy KM, Dawson JC, Jones SS (2008) Relationship among phenotypic growth traits, yield and weed suppression in spring wheat landraces and modern cultivars. Field Crops Res 105:107–115

Murray DL (1982) The abolition of El Cortito, the short handled hoe: a case study in social conflict and state policy in California agriculture. Soc Probl 30:26–39

Owen MDK, Zelaya IA (2005) Herbicide-resistant crops and weed resistance to herbicides. Pest Manag Sci 61:301–311

Pimentel D, Hepperly P, Hanson J, Donds D, Seidel R (2005) Environmental, energetic and economic comparisons of organic and conventional farming systems. BioScience 55:573–582

Refsgaard K, Halberg N, Kristensen ES (1998) Energy utilization in crop and diary production in organic and conventional livestock production systems. Agr Syst 57:599–630

Reganold JP, Glover JD, Andrew PK, Hinman HR (2001) Sustainability of three apple production systems. Nature 410:926–930

Sadek NF, Oktarani YP (2009) Consumer knowledge and perception about organic food: a challenge for consumer education and the benefits of going organic. Asian J Food Agro-Ind, Special Issue S363–367

Smil V (2001) Feeding the world: a challenge for the twenty first century. MIT Press, Cambridge, MA

Stolz H, Stolze M, Hamm U, Janssen M, Ruto E (2011) Consumer attitudes towards organic versus conventional food with specific quality attributes. Wagening J Life Sci 58:67–72

Swanton CJ, Harker KN, Anderson RL (1993) Crop losses due to weeds in Canada. Weed Technol 7:537–542

Swinton SM, Buhler DD, Forecella F, Gunsolus JL, King RP (1994) Estimation of crop yield loss due to interference by multiple weed species. Weed Sci 42:103–109

Thompson G (2000) International consumer demand for organic foods. HortTechnology 10:663–674

Timmons FL (1970) A history of weed control in the United States and Canada. Weed Sci 18:294–307

USDA, ERS (2011) Economic Research Service. http://www.ers.usda.gov/data-products/adoption-of-genetically-engineered-crops-in-the-us/recent-trends-in-ge-adoption.aspx#.Ubi95ticGVI. Accessed 6 June 2013

van der Schans D, Bleeker P (eds) (2006) Practical weed control in arable farming and outdoor vegetable cultivation without chemicals, Publication number 352. Applied Plant Research, Wageningen

Vecchia AV, Gilliom RJ, Sullivan DJ, Lorenz DL, Martin JD (2009) Trends in concentrations and use of agricultural herbicides for Corn Belt Rivers, 1996–2006. Environ Sci Technol 43:9096–9102

Walz E (1999) Final results of the third biennial national organic farmers' survey. Organic Farming Research Foundation, Santa Cruz

Willer H, Kilcher L (eds) (2011) The world of organic agriculture. Statistics and emerging trends 2012. FIBL-IFOAM report. Research Institute of Organic Agriculture (FiBL), Frick and International Federation of Organic Agriculture Movements (IFOAM), Bonn

Willer H, Kilcher L (eds) (2012) The world of organic agriculture. Statistics and emerging trends 2012. FIBL-IFOAM report. Research Institute of Organic Agriculture (FiBL), Frick and International Federation of Organic Agriculture Movements (IFOAM), Bonn

Williams MM II, Boydston RA (2005) Alternative to hand-weeding volunteer potato (*Solanum tuberosum*) in carrot (*Daucus carota*). Weed Technol 19:1050–1055

Williams MM II, Ransom CV, Thompson WM (2007) Volunteer potato density influences critical time of weed removal in bulb onion. Weed Technol 21:136–140

Yim FS, Williams MM II, Pataky JK, Davis AS (2009) Principal canopy factors of sweet corn and relationships to competitive ability with wild-proso millet (*Panicum miliaceum*). Weed Sci 57:296–303

Zimdahl RL (2000) My view. Weed Sci 48:663

Section II
Principles and Merging of Engineering and Weed Science

Chapter 3
Engineering Advancements

John K. Schueller

Abstract Significant advances have recently occurred, and are continuing to occur, in many of the major engineering fields. Improved motions, structures, and their control are seen with contemporary mechanical engineering technologies. New materials and manufacturing processes make it practical to substantially improve machine performances. Significant advances in computational speed, storage capacities, and networking have made electronic and electrical components, such as machine vision and spectral sensors, more practical and productive. Automation and control technological advances have greatly improved potential system performances. These are among the many engineering advancements which provide techniques to facilitate the technical and economic feasibility of automated weed control.

1 Introduction

These are exciting times for most engineering fields. Almost all fields are seeing significant advances in research and technology development. Since these advances may impact the technical and economic feasibility of automated weed control, this chapter will review developments in a few of those engineering fields.

J.K. Schueller (✉)
Center for Manufacturing Innovation, University of Florida,
Gainesville, FL 32611-6300, USA
e-mail: schuejk@ufl.edu

S.L. Young and F.J. Pierce (eds.), *Automation: The Future of Weed Control in Cropping Systems*, DOI 10.1007/978-94-007-7512-1_3,

2 Mechanical Engineering

Mechanical engineering is perhaps the most diverse field of engineering. Many technologies are used and they are applied in many industries. This wide range of applications has led to it currently being the largest field in numbers of engineering students. This is, of course, also reflected in the wide diversity of subfields in which there is significant activity. There is not space here to review even all the major subfields. So the focus will be on some of the subfields which are the most rapidly advancing.

2.1 Kinematics and Dynamics

The motion of mechanical elements to perform tasks depends upon accurate and fast movements. These, often robotic, movements need to be accurately controlled. Kinematic analyses determine the paths of mechanical components. Dynamic analyses include the effects of mass, damper, and spring elements in studying these motions. Motion analyses are used to study known motions and motion synthesis is used to get desired motions.

Kinematic and dynamic analysis and synthesis continues to be an area of fundamental study in mechanical engineering. Since the equations are usually nonlinear, this remains a difficult area. But progress continues to be made. For practical implementations, simulation computer software packages are developed and widely available. They allow users to simulate mechanical device motions. The constantly improving computation abilities allow more rapid studies of motions. Optimization and other advanced techniques can be used to find the optimum geometries of mechanical components, thereby significantly improving speed and accuracy of the tasks being performed.

Most robotic or automated mechanisms are serial mechanisms in which there are a series of links connecting sequential mechanical elements. Forward kinematics seeks to find the positions of the mechanism components from the joint angles between links and the link extensions. The mathematics of these calculations using matrix algebra is now well known. Reverse kinematics is the more difficult task of finding the angles and extensions from the desired positions. Substantial progress has been made in both forward and reverse kinematics so that it is now possible to design the motions of mechanical components well.

Parallel mechanisms are now becoming more popular (right half of Fig. 3.1). The kinematics and dynamics of such mechanisms are becoming more well known, although they are substantially more complicated. Again, the advances in computational performance make the solving of the mathematics of such systems easier.

2.2 Automotive Engineering

The automobile industry has the greatest role of any mechanical industry in many economies. Many technologies, or the widespread commercialization of technologies at reasonable prices, come from the automobile industry. It has had an even

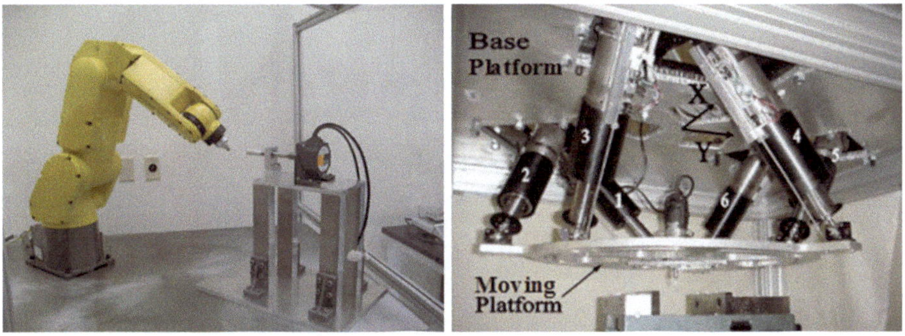

Fig. 3.1 Example serial (*left*) and parallel (*right*) manipulators (Photos courtesy of H.Y. Greenslet and G.J. Wiens)

greater than expected effect on the agricultural equipment industry because it shares personnel (such as Henry Ford and the United Auto Workers union), components (e.g., engines), and manufacturing techniques.

One trend in the contemporary automobile industry is the adoption of alternative powertrains. Instead of simple gasoline or diesel engines, the industry now has parallel and serial hybrid drives which consist of engine and electric motor elements. The drive for energy efficiency has pushed the use of electric motors in power applications and the use of regeneration and energy recovery schemes in braking and other procedures. Some plug-in electrics dispense with the engine completely.

Power transmission research and development has regained importance in the drive to increase efficiency. Transmissions have many more gears and alternative structures, such as dual-clutch arrangements. Continuously-variable transmissions can exactly match the performance demand speeds of the loads to the maximum efficiency speeds of the power supply.

To achieve optimum performance and maximum operator comfort, controllers are used throughout modern automobiles. An upscale automobile may have as many as 100 controllers, many networked together. This diverse and numerous controller use is spreading beyond automobiles to other industries.

2.3 *Autonomous Vehicles*

One of the most exciting areas of mechanical engineering research has been the development of autonomous vehicles. There have been many research and development projects in land, sea surface, underwater, and airborne environments. Most of the autonomous vehicle work has been done with ground vehicles, including some substantial efforts with agricultural vehicles.

The most-publicized activities, involving hundreds of independent efforts, were the DARPA Grand Challenge (in 2004 and 2005) and Urban Challenge (in 2007) prize contests sponsored by the US Department of Defense. These robotic development contests significantly advanced the state of the art and created a large pool of

Fig. 3.2 Autonomous orchard sprayer

robotics experts by challenging groups from universities and the private sector to develop vehicles which could successfully travel and perform tasks. The successful autonomous vehicles demonstrated the importance of path planning, sensing, and robust automatic behaviors. Many technologies and techniques were substantially further developed.

Perhaps agricultural autonomous vehicles can be classified into categories of input-output, transport, and self-contained. Many aspects of agricultural production can be considered a materials handling problem. In that manner, there is a big logistics problem in getting materials (such as seeds and fertilizers) to the fields and harvested crops from the fields. The actual input (planting or transplanting) and output (harvesting) operations within the field are crucial, complex operations. They will perhaps be the last operations to be autonomous. However, given the complexity, they tend to already have automated functions on the machines. On the other hand, transporting the materials to and from the field might soon utilize autonomous vehicles if the interface with the input-output vehicles continues to improve. Examples of self-contained vehicles are those which perform scouting, tillage, weeding, or spraying (Fig. 3.2). They function by themselves and the technology is rapidly developing.

Similar advances have been made in small aerial vehicles often called UAVs or drones. Small airplanes, helicopters (Fig. 3.3), multicopters, blimps, and other flying vehicles have become more widely researched and used. Advances in sensing and controls have made them easier to use and more robust. They have great potential in agriculture. Payloads are limited, but the overhead perspective is useful,

Fig. 3.3 Small traditional helicopter used for scouting and yield mapping

especially for sensing, scouting, and management tasks. Vehicles with four or eight rotors are easier to control than traditional helicopters with only one lift rotor. The problem of short aerial persistence is being alleviated by the development of autonomous refueling or battery recharging stations.

2.4 Surface Science

Recent mechanical engineering advances have also occurred in the area of surface science. Surfaces are very important because they affect the interaction between items. With advances in understanding and measuring the small scales which affect surface performance, better surfaces are now being created.

Wear and friction are now better understood. Although far from perfect, better predictions can now be made. These predictions and improved manufacturing techniques can be used to produce better surfaces. A wide variety of materials can be applied to surfaces. Many new techniques, including various powder and plasma techniques, can now be utilized to apply materials with specific content and surface characteristics.

In addition, the shapes of the surfaces can be modified or created for specific functional characteristics. Magnetic-assisted finishing can control surface characteristics, as can techniques borrowed from the microelectronics and nanotechnology industries. For example, it is now known how to make surfaces which attract or reject certain materials or biological entities. One such example is hydrophilic and hydrophobic surfaces, which attract and repel water respectively (Fig. 3.4).

3 Materials and Manufacturing in Mechanical Engineering

All components of agricultural equipment need to be made of materials and manufactured. Traditionally, agricultural equipment has been produced with materials and manufacturing methods which were economical, but not sophisticated compared to other industries. However, increasing performance demands are increasing the requirements of materials and manufacturing for contemporary equipment.

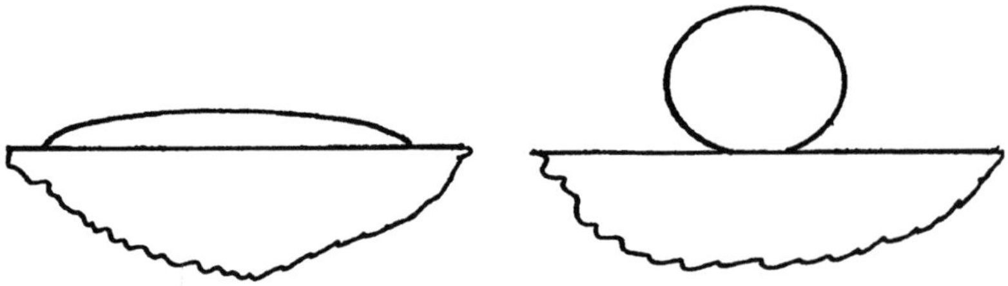

Fig. 3.4 Examples of droplet behaviors on hydrophilic (*left*) and hydrophobic (*right*) surfaces

3.1 Lightweight Metals

Like other cost-sensitive heavy mechanical industries, the agricultural equipment industry has traditionally used steels and cast irons. There has been some use of lightweight metals, but the untapped potential is for much more usage. This corresponds with technology developments in processing lightweight materials and their increasing usage in other industries.

Aluminum alloy usage continues to expand. With modern cutting tools, aluminum alloys can be machined at great speeds on contemporary equipment. Good welding techniques have been developed and popularized. Die casting and extrusion have become efficient mass production processes. Therefore, the technology is present to substitute lightweight alloys to improve performance if the proper aluminum alloys and heat treatments are selected.

Other lightweight materials are also becoming more feasible and used for widespread applications. Magnesium currently seems to have the most interest. Magnesium alloys are even lighter than aluminum alloys and can be machined and cast similarly. Care must be taken with some of these metals, especially if small chips or fines are generated, to avoid fires (e.g., from magnesium) or toxicity (e.g., from beryllium which is lighter and stiffer than magnesium).

3.2 Composites

There is even more progress and excitement in the development and commercialization of composites. Composite structures are now achieving market penetration in aerospace, automotive, construction, and other industries due to advances in the composite materials, techniques for design of composite structures, and better composite manufacturing. A composite is technically any material made from two or more different constituent materials with different properties. Although that definition includes such items as concrete and wood, the term is usually used to refer to man-made materials which have a reinforcement (often a fiber) and a matrix. The fibers are usually strong and carry the load, but are brittle. The matrix provides toughness, transfers the load to the fibers, and protects them. In this way, a strong, lightweight component is produced without too much brittleness.

Composite materials can be designed to have specific properties by the choice of the component materials and their proportions. For example, carbon fiber in a polymer matrix is a common combination and is used in components from car parts to golf clubs. Composites can be designed to have specific weights, stiffnesses, strengths, etc. In addition, the properties can be anisotropic (different in different directions) and can be varied throughout a component (functionally graded) to optimize cost and performance.

Composite manufacturing methods are being improved and being made more economical. Automation and robotics have removed some of the labor requirements which have often made manufacturing with composites financially noncompetitive. As composites are getting wider use in various other industries, they are becoming more practical for agricultural applications where light weights are crucial. Lightweight composite components in mechanisms can be rapidly accelerated and moved.

3.3 Rapid Prototyping

Rapid prototyping, also known as additive manufacturing, free-form manufacturing, or 3D printing, is revolutionizing manufacturing, especially for producing single items or small quantities. Many different rapid prototyping technologies, under many different names, have been developed. These technologies generally use a computer-controlled system to build up a mechanical component layer by layer. Complex geometries are automatically created by the computer control of the process.

There are a large variety of rapid prototyping technologies. Generally, a computer-aided design program is first used to create a CAD model of the component. The model is then divided into thin slices. The rapid prototyping machine then builds the components slice by slice. Common technologies include:

- Stereolithography: photosensitive liquid polymer is solidified by laser light.
- Selective laser sintering: laser beam fuses powdered material (usually metals).
- Laminated object manufacturing: perforated sheets of solid material are bonded together.
- Fused deposition modeling: filaments of heated thermoplastic are extruded from a nozzle.
- 3D ink-jet printing: ink-jet technology is used to print layers of powder glue or plastics.

After the objects have been formed, they are often cleaned and further processed.

Rapid prototyping allows the quick construction of complex components directly from computer files. It found its initial use in prototyping to decrease product development time and facilitate communication within organizations designing and producing products. But it has now expanded to also producing tools (such as dies, molds, and fixtures) and actual products. Rapid prototyping may make the development of custom components for agricultural equipment easier. It may also allow the easy construction of replacement parts in the farm shop. In addition, the technology advances in high-speed laser and nozzle manipulation in these manufacturing machines might be very applicable to equipment designed for other purposes.

3.4 High-Speed Actuation

Even beyond rapid prototyping, there has been a continued revolution in high-speed actuation in manufacturing. Linear motors provide drives with very high-speed movement and rapid acceleration. They have become popular in EDM (electrical discharge machining) and other machine tools. Both linear electrical drives and traditional rotary electrical drives have greatly improved their performance through being designed by engineers with a better understanding of structural dynamics, and the resulting needed filtering and compensation. Digital controls have allowed unwanted resonances to be removed.

3.5 Machine Structures

Manufacturing machines, such as machine tools, traditionally have had compounded axes. The state of the art in designing such axes has evolved. In addition, there has been a big trend to move from three-axis to five-axis machines. The additional axes give more flexibility and allow more complicated motions to be performed. New designs of machines also include multifunction machines which can mill and turn on the same machine and machines which use parallel kinematic structures instead of serial kinematic structures. The practical commercialization of these designs has implications for nonmanufacturing machines. It is now easier to have machines with nontraditional structures and with powered components that can be moved on complex paths.

The more axes and new designs greatly complicate path planning and control. However, advances in understanding, and especially the great increases in computational capabilities, allow the machines to be effectively utilized. The use of accurate and fast control is now possible to generate complex motions with complex machines.

3.6 Cutting Processes

Although additive processes are becoming more important and net-shape and near-net-shape processes (such as casting, extrusion, and forging) are continuously being improved, cutting still remains the most important category or manufacturing processes.

Raw materials are usually made into finished components through turning, milling, drilling, threading, and other processes. The productivity and cost of such processes continue to improve due to improvements in cutting speeds. Besides the advances in structures, actuators, and controls discussed above, there has been a very substantial improvement in cutting tools. Although the geometry of the tools has been improved, the biggest productivity gain has been due to the development of advanced cutting tool materials. Most cutting is done with cemented tungsten carbides. Finer grain sizes and complex compositions, combined with multiple layers of chemical and physical vapor deposition (CVD and PVD) coatings, have made strong, tough, and heat-resistant tools commonplace.

The advances in cutting tools are applicable to agricultural tools which cut plants and soil. For applications where solid carbides are not appropriate, carbide coatings have a huge potential. The use of high velocity oxy-fuel and high velocity air fuel (HVOF and HVAF) thermal spray technologies is becoming widely commercialized. Components with these coatings can be strong and wear resistant.

3.7 Nanotechnologies

One of the biggest areas of recent research and development has been in the various nanotechnologies. This is the study, manipulation, and application of materials and components at the nanometer scale. This study has occurred in the fundamental biological, chemical, and physical sciences, as well as more applied fields such as agriculture, medicine, and engineering. Techniques and instrumentation have been developed to work at these small scales. Besides allowing the manipulation and construction of materials at near-atomic levels, the techniques and instrumentation developed to work at these small scales will help permit a better understanding of large-scale behaviors. For example, it may allow a better understanding of spectral reflectivity and other characteristics for sensor development and other uses.

One of the achievements of the increased nano capabilities is the manufacture and use of nanoparticles and other nanostructured materials. For example, there is a possibility of embedding unique nanoparticles within crop seeds and plants. Nanoparticles and nanostructured materials are being used to produce strengthened and otherwise improved materials. They are also being used to improve sensors.

The ability to actuate and deform at very small scales has opened many opportunities to have materials with unique properties, such as the surfaces discussed above. In fact, it is now possible to create nano, or micro, devices. As the technologies are rather new, it is unclear how they might be applied in the large-scale world we live in. The devices are small, but they might be used in multiscale devices. For example, nano or micro elements might form the first stage in multistage sensing or control. Such a stage may perform secondary control on a higher control stage. Micro or nano devices also may form the vital recognition and identification stages in sensors (Fig. 3.5).

Fig. 3.5 Scanning electron microscope image of tool-tip gripping mechanism (Photo courtesy of C. Taylor)

4 Electronics and Electrical

The capabilities and uses of electricity and electronics continue the increasing importance they saw throughout the twentieth century. Electric and electronic components are increasingly commonplace. Although advances in traditional electrical and electronic subfields (such as motors or amplifiers) continue, perhaps the greatest revolutions are occurring in other subfields.

4.1 Microelectronics

The proliferation of electronics in contemporary society was spurred by the developments in designing and manufacturing microelectronics. Sophisticated, powerful devices are now available at very low costs. Moore's Law (which predicts that computer performance doubles in less than every 2 years) and similar relationships have increased power while reducing costs.

Calculation speeds continue to increase as a result. More sophisticated algorithms, even computationally intensive algorithms such as sound, image, or video processing, can now be rapidly run. Information from sensors can be more complex and can be generated much more quickly. Real-time processing can occur in which sensor data can be processed and the desired control changes made immediately. Storage costs have similarly decreased. It is now possible to track every crop plant and to store voluminous history data on each plant. Economical storage is no longer a limit.

4.2 Networking

A variety of concepts, such as ubiquitous computing, pervasive computing, Internet of Things, and cloud computing, refer to various trends in having computers everywhere and connected together. Information, computing power, and knowledge transfer are becoming omnipresent. Information can be processed and shared between large databases (such as detailed farm records on the cloud), fixed infrastructure (such as an irrigation network), and mobile machines (such as tractors and other farm equipment).

Networks share and transfer the information. The Internet provides a capability for the individual farming operation to communicate and share information with suppliers and other agribusiness concerns. Information can be warehoused in an off-farm location given the expansion of networking and cheap data storage.

There has been significant progress in wireless networking. High-performance, reliable, and affordable wireless systems are widely available. Connections to mobile equipment and dispersed fixed infrastructure can now be made with wirelessly at affordable costs and now for the distances commonly required in agriculture.

This networking allows communications to mobile agricultural equipment. It also allows controllers on a piece of equipment to communicate to each other, thereby improving the overall agricultural system's inherent performance. Within a single piece of equipment, the networking can be either wired or wireless.

4.3 Machine Vision

Machine vision sensors are one of the items which can be networked. Machine vision was an area of much interest and progress in the 1980s. However, its commercial success was uneven due to sensor and computational processing limitations. These limitations have largely been removed. Now even cell phones have cheap vision sensors and processing is fast and inexpensive as discussed above.

As seen in both still and video cameras, vision sensors have achieved better performance at lower costs. The resolutions and speeds have increased. The sensors are now able to respond to lower light levels.

Researchers have also developed more productive and robust algorithms for such tasks as filtering and object identification. Commercial development is being led by advances in many industries, including for automotive and defense applications. Machine vision techniques can identify items and their locations in various environments and conditions.

4.4 Spectral Sensing

Spectral sensing is related to machine vision and is sometimes integrated as part of a machine vision process. Spectral reflectance can be sensed to identify objects or an object's characteristics. For example, it can determine or help determine a species of a plant. Or it can determine the physical or chemical characteristics of the reflecting surface, such as the nutritional status of a plant.

Advances in hardware have extended the range of frequencies of the electromagnetic spectrum which can be used for sensing. Historically, portions of the visible and near-infrared spectrum were used for agricultural sensing. Now, technologies are being developed to utilize other portions, such as far-infrared, ultraviolet, and terahertz frequencies. Sensors are not only becoming more economical, but they are also utilizing narrower bands and more channels. Increasing computational speeds allow more sophisticated and complicated combinations of spectral frequency responses to be used to achieve improved performance.

5 Automation and Computer Science

The advances discussed above in the electronic and electromechanical fields have been greatly facilitated by advances in electronic hardware. However, there have also been software advances. Some of them are discussed above. In addition, there has been a general increase in programming reliability and performance, improving automation performance.

5.1 Control Theory

Control theory continues its development to improve the dynamic performance of systems to which it is applied appropriately. Classical control theory using the frequency domain and Laplace transform analysis is now well understood. Systems can be designed for good steady-state and transient response and yet have stability. This results in faster performance.

Modern control theory continues to gain popularity relative to classical control. Modern control describes the system in the time domain with a set of first-order ordinary differential equations, usually expressed in matrix form. It is often better at handling controller synthesis and multivariate applications, although it is less intuitive to many controls designers and users. Pole placement techniques can be used to get desired system responses.

A wide range of new techniques, such as fuzzy control, neural networks, and genetic algorithms, have been developed. These biologically inspired nonlinear techniques can often improve performance or robustness if they are configured properly.

Although they generally do not have strong mathematical provability, they can exhibit adequate behavior in practice.

A very general and widespread trend in automation is the tendency for the human to no longer be the direct controller. Direct control is increasingly relegated to automated systems with the humans moving to higher, more abstract levels. The human is the supervisor, not part of the actuation system.

5.2 Secondary Control

If a hydraulic or pneumatic mechanism is to be actuated, there must be some way of causing the mechanism to move or not move according to some desired path as a function of time. This control is usually traditionally achieved by having a valve to modulate the flow of oil (hydraulics) or air (pneumatics) to the cylinder or motor which is performing the actuation. (This is similar to how an electromechanical actuator will have some electronic or electrical components to control the current and voltage to it.) This valving tends to produce systems which respond quickly and have accurate controllability. However, there are often substantial pressure, and hence energy, losses in regulating or throttling the flow through the valve.

Consequently, there has been an increasing interest in secondary control and related techniques. Instead of directly controlling the actuator, the control is more remote. For example, instead of using a valve to control a hydraulic cylinder, the pump providing the hydraulic oil itself is controlled. There is a much higher efficiency in the system since excess power is not generated and dissipated. However, since the control is further from the actuator, there is often difficulty getting fast and accurate control. In addition, more power sources may be required when there are multiple actuators, as shown by examples of secondary control of excavators. However, recent progress now makes secondary control feasible in an increasing number of cases.

There has also been progress in the regenerative capturing of energy during deceleration. Instead of dissipating energy into heat through friction, restricting flow, or resistance heating, energy is stored for later productive use. One example is the use of regenerative braking in hybrid automobiles. Similar concepts are possible in other systems through the use of accumulators, batteries, ultracapacitors, and other energy storage devices.

5.3 Formation Control

The control of multiple autonomous machines has been an area of significant study and development. This formation control allows many separate machines to be controlled and to perform a task which would be impossible for a single small machine to perform, or to perform with acceptable accuracy or productivity. An implication of the ability to control formations is that smaller, multiple machines may replace large machines without requiring multiple human operators. These machines can be smaller, and perhaps be more efficient, have lower environmental impacts, and be less hazardous. In addition, the natural redundancy of multiple machines may increase system reliability.

5.4 Human-Machine Interaction

HMI, whether representing human-machine interaction or human-machine interface, is becoming more important as there are more machines to interact with. And the HMI is happening at higher levels of abstraction. Presentation of information to the human should utilize the improved human factors knowledge developed through the utilization of psychological and ergonomic research.

The best methods for human interface design are changing rapidly as different generations of humans are maturing and the knowledge and habits from their different experiences change. Visual, sound, and kinematic outputs and inputs should be designed to maximize robust performance. Display and control stereotypes continue to evolve.

5.5 Computer Science

As computational capabilities greatly increase in both hardware and software, the importance of computational efficiency relatively decreases. Computer science is moving away from electrical engineering and becoming more concerned with human issues. Although the vast databases and networking demand efficiencies, the emphasis is now on higher levels of abstraction and human interfaces.

As computer programs become more complex, the substantial shift from procedural programming to object-oriented programming continues. This is seen by the widespread use of languages such as C++ and Java and by the use of interactive software environments such as MATLAB or LabVIEW. As more computer programs, packages, and tools have been developed, programming increasingly involves reusing existing code or constructs.

The software engineering management of computer programming and how programs are written has also changed. A process called scrum is now often used for the actual writing of software. Scrum is a holistic product development strategy in which the development team works as a unit to reach a common goal rather than an individualized sequential approach. The team works in short time-limited sprints in which portions of the system are created. The process emphasizes self-organizing co-located teams with good verbal communication among themselves.

6 Conclusion

There continue to be advances in many engineering fields. This chapter lists just a few of the trends. The following chapters show how some of these trends are impacting the potential for successful commercial development of automated weed control. Ultimately, it will be clever scientists and engineers who will find many new ways in which these advances can be applied to the field.

References

Cox SWR, Schmoldt DL (eds) (2001) Past developments and future directions. Comput Electron Agric 30(1–3):1–246

de Silva CW (2010) Mechatronics: a foundation course. CRC Press, Boca Raton

Krutz GW, Schueller JK (2000) Advanced engineering: future directions for the agricultural and biological engineering profession. J Agric Eng Res 76:251–265

Oliver MA, Bishop TFA, Marchant BP (2013) Precision agriculture for sustainability and environmental protection. Routledge, London

Rovira-Mas F (ed) (2012) Automation technology for off-road equipment 2012. In: Proceedings of the 10–11 July conference, Valencia

Schueller JK (1992) A review and integrating analysis of spatially-variable control of crop production. Fertil Res 33:1–34

Schueller JK (2000) In the service of abundance: agricultural mechanization provided the nourishment for the 20th century's extraordinary growth. Mech Eng 122(8):58–65

Zhang Q, Pierce FJ (eds) (2013) Agricultural automation: fundamentals and practices. Taylor & Francis Group, Boca Raton

Chapter 4
Plant Morphology and the Critical Period of Weed Control

J. Anita Dille

Abstract The critical period of weed control (CPWC) provides a time frame in the life cycle of the crop for scouting or sensing weed populations and making weed control applications to prevent crop yield losses. This time frame is relatively early in the growing season for a given crop. Thus, newly emerged and small weed seedlings need to be observed prior to the start of the CPWC. Morphological characteristics of these seedlings are diverse and influence the ability to sense the seedlings at a given time, and account for the changes that have occurred over time. Understanding the population dynamics for different weed species, that is, emergence timing relative to the crop, types of cotyledons and leaf arrangements, and rates of leaf appearance and stem elongation, better scouting or sensing methods can be designed prior to the start of the CPWC. The development of new automated technologies must take into account changing morphology of weed seedlings early in the life cycle of the crop. Crop management practices influence the types and numbers of weed species present and thus can change the beginning and end of the CPWC and the timing for scouting and control.

1 Introduction

When developing weed control strategies, knowledge of plant biology is one of the most important factors. Without an understanding of the changes that occur during plant growth and development, most weed control practices will result in less than satisfactory control. This chapter will discuss the basics of weeds in cropping

J.A. Dille (✉)
Department of Agronomy, 3701 Throckmorton Plant Sciences Center,
Kansas State University, Manhattan, KS 66506, USA
e-mail: dieleman@ksu.edu

S.L. Young and F.J. Pierce (eds.), *Automation: The Future of Weed Control in Cropping Systems*, DOI 10.1007/978-94-007-7512-1_4,

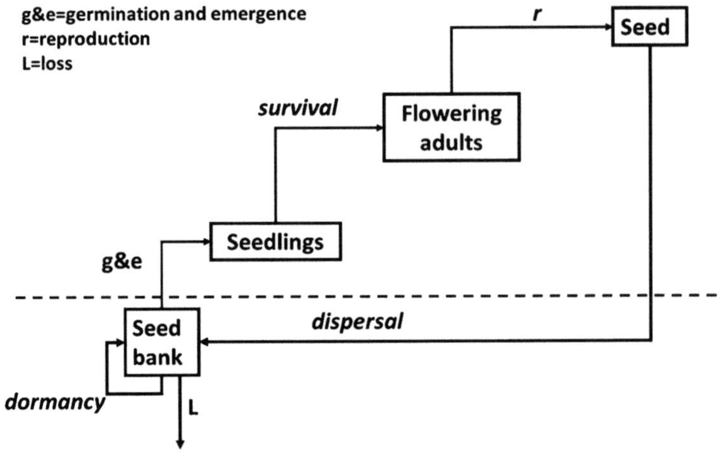

Fig. 4.1 Population dynamics of an annual weed species showing stages (*boxes*) and transitions (*arrows*) among stages. *Dashed line* indicates above and below soil surface

systems from the perspective of a weed scientist. The focus will be on weed morphology and the timing for implementation of critical weed control measures. In addition, the importance of an integrated weed management (IWM) approach will be emphasized.

Life cycles of weed species are described as summer annuals, winter annuals, biennials, or perennials (Ross and Lembi 2007). Summer annuals will emerge in the spring of the year and grow through the summer until fall when they produce seed and die, while winter annuals tend to emerge in the fall and overwinter to flower, set seed, and die in the spring. Biennials require 2 years to complete their life cycle from seedling to new seed, and perennials live longer than 2 years.

Population dynamics of a weed species are described using a series of stages and transitions that a plant goes through during a year. Stages are those observable conditions of a plant such as seed, seedling, and flowering plant, while transitions are rates of germination and emergence, proportion of seedlings that survive to become flowering plants, and rate of reproduction by flowering plants to produce new seed (Fig. 4.1). With annual weed species, the seed is the key stage in order for the population to perpetuate into the following year. Anderson (2005) highlighted three stages and transitions to target for weed control, that is, (1) enhancing the natural loss (L) of weed seeds in the soil seed bank, (2) reducing weed seedling establishment (survival), and (3) minimizing seed production by individual plants that survive to maturity (r) (Fig. 4.1).

Each stage in the population dynamics cycle can be influenced with one or more weed control tactics as part of an IWM strategy. Tactics include cultural, biological, mechanical, and chemical practices. For example, cultural tactics include rotation design and crop sequencing, use of no-tillage practices, crop residue management, and developing competitive crop canopies via fertilization strategies and row spacing/seeding rate adjustments (Anderson 2005; Swanton and Weise 1991).

Mechanical or physical control tactics include any preplant or in-crop tillage practice, flaming, or use of mulches to physically reduce weed seedling establishment and growth. Chemical practices include application of preemergence and postemergence herbicide products that can provide residual, contact, or systemic activity. Many of these tactics can be implemented using automated technologies that might provide for more accurate placement of a control practice while reducing environmental and economic impacts of weed control.

Within the framework of the weed population dynamics cycle, weed control is often implemented to reduce weed seedling establishment either as seed germinates and emerges (g&e) or just after emergence (survival) (Fig. 4.1). The type of knowledge needed to effectively implement weed control practices varies depending on ecological, economical, or efficacy-based perspectives. From an ecologically based perspective, knowledge is needed on weed species germination and emergence timing (relative to the crop), on weed seedling growth rate including the production of leaves and elongation of stems within a given crop, and on impacts of a given crop production system on weed seedling growth and development. From an economically based perspective, it is important to know the potential yield loss, reduction in crop quality, or impact of future weed seed production as a result of those weeds. A combination of ecology and economics determines the critical period of weed control (CPWC) which is defined as the time period in the crop growth cycle when weeds must be controlled to protect future crop yield (Knezevic et al. 2002). The CPWC is useful as a time frame for making decisions on the scouting for weeds and timing of application of weed control practices. From an efficacy-based perspective, knowledge is required on the maximum size when weed seedlings can still be effectively controlled with a given weed control tactic. For example, postemergence herbicides work more effectively on small, actively growing weed seedlings, and maximum size limitations are often provided on herbicide labels. The use of mechanical tools such as a rotary hoe requires weed seedlings to be just emerging through the soil surface, or the weeds will be too large for effective control with this tool. Weed seedlings can be somewhat larger if using an inter-row cultivator. Other crop production practices that influence weed emergence and how weeds change through time include crop species present, row spacing and crop seeding rates, fertility practices and tillage systems, and any previous weed control practices implemented. Depending on choice of weed control practice, one needs to be able to identify or "sense" the weed species before it reaches this maximum size and needs to be able to effectively control these plants, preferably during the CPWC.

The overall goal of this chapter is to highlight the challenge of observing weeds at a resolution relevant for both automated weed control (sensing and control) and in a timely manner with respect to minimizing impacts on crop growth and development that result in yield losses. Key concepts will be (1) when do weeds emerge, (2) what do they look like at emergence and as they grow, (3) how quickly do canopy features of weed species change through time, (4) when do these weeds impact crop yield as described using the critical period of weed control, and (5) what is the influence of crop production practices on weed emergence and growth.

2 Weed Emergence Timing

Within an agricultural field, it is the norm rather than the exception to have some weed species emerging early and others later in the growing season. There is a continuum of emergence that can be described, such as when a weed species begins to emerge, how rapidly individual seedlings appear, and for how long they continue to emerge during the growing season. Emergence patterns can be determined through regular weekly census of naturally occurring weed populations. For example, four common summer annual weeds in corn and soybean fields in Iowa had velvetleaf (*Abutilon theophrasti*) emerging the earliest, while common waterhemp (*Amaranthus rudis*) emerged later, and it had a longer emergence period than woolly cupgrass (*Eriochloa villosa*), giant foxtail (*Setaria faberi*), or velvetleaf (Hartzler et al. 1999). The pattern of kochia (*Kochia scoparia*) seedling emergence in the central Great Plains was primarily from early April to late June, whereas green foxtail (*Setaria viridis*), wild-proso millet (*Panicum miliaceum*), and redroot pigweed (*Amaranthus retroflexus*) began emerging in late May and continued until August (Anderson and Nielsen 1996).

The rate of emergence varies by species and can occur in single or multiple peaks. Some weed species, such as kochia, will have a majority of seedlings emerging within 7 days of initiation (Dille et al. 2012), while other weed species, such as common waterhemp, will have a prolonged time period over which seedlings appear (Hartzler et al. 1999). Woolly cupgrass consistently had its initial emergence later than many species, but nearly all seedlings occurred within 3 weeks after initiation. For other weeds, such as velvetleaf, giant foxtail, and common waterhemp, emergence peaks are typically influenced by rainfall events (Hartzler et al. 1999). Depending on the year and related rainfall events, the total number of weed seedlings that will emerge varies, but typically, the initial emergence for a given weed species begins at approximately the same time each year.

When describing a weed community made up of many species, there may be several emergence peaks as well. For example, kochia and Russian thistle (*Salsola tragus*) were the main species emerging during the first peak early in the year, whereas green foxtail, wild-proso millet, and pigweed species (*Amaranthus* spp.) predominated in the second peak that occurred later in the growing season (Fig. 4.2). In this example, the two peaks represented a majority (67 %) of the total weed seedling emergence. The emergence pattern of this weed community was consistent among years, and crop rotation or tillage system did not change the pattern, only total number of seedlings that emerged (Anderson 1994).

Knowing the seedling emergence pattern of a weed population or community provides insight as to what crop should be planted and options for weed control. For example, Anderson (2005) overlaid the emergence pattern of the weed community in the Great Plains with the timing of corn or sunflower planting (Fig. 4.3). Corn is normally planted in early May, whereas sunflower is planted 3–4 weeks later. This delay with planting provides producers with an additional opportunity to control 35–50 % of potential weed seedlings before planting sunflower; otherwise, these weed seedlings emerge in corn and require post-planting control. In another example in

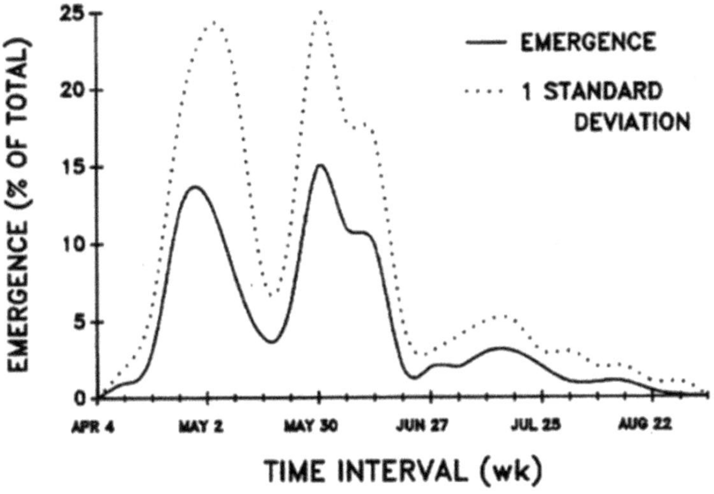

Fig. 4.2 Seedling emergence pattern of a weed community (*solid line*) at Akron, CO averaged over 7 year 1987–1993. *Dotted line* represents 1 standard deviation (Reprinted with permission from Anderson 1994)

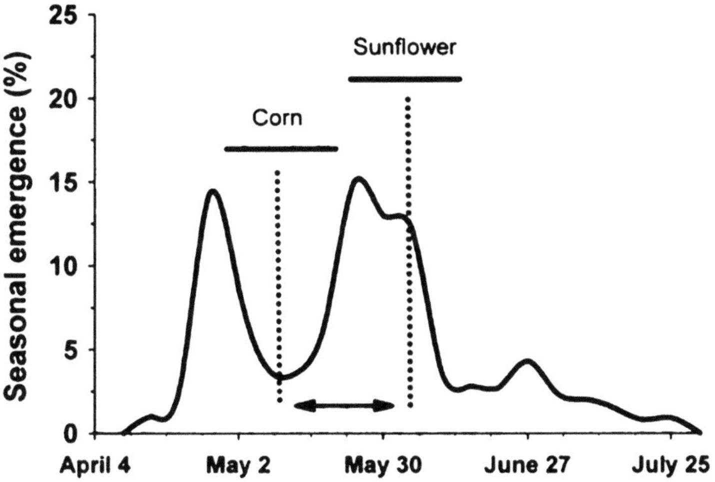

Fig. 4.3 Seedling emergence pattern of a weed community at Akron, CO. Data averaged across 7 year. *Horizontal lines* underneath corn and sunflower represent normal planting dates for these crops. The *double-ended arrow* highlights the potential difference in seedling emergence between average planting dates of these crops (Reprinted with Permission from Anderson 2005)

Iowa, optimal corn planting dates are between April 15 and May 9 (Elmore 2012), and optimal soybean planting dates are after April 25 (Pedersen 2013). The following weed species are problems in these two crops because the initial emergence of velvetleaf is on April 28, followed by woolly cupgrass on May 2, giant foxtail on May

15, and common waterhemp on May 22 in Iowa (Hartzler et al. 1999). These data highlight the need to know when different weed species emerge in relation to the crop being planted.

3 Weed Emergence and Growth Characteristics

As weed species germinate and emerge from the soil, the first plant part that is exposed above the soil surface is usually the cotyledons. This immediately impacts that amount of green plant material that could be observed and sensed. If the plant is a flowering annual weed species, it could be a dicotyledon with two cotyledons (broadleaf) or it could be a monocotyledon with one cotyledon (grass or grasslike). Many weed species can be easily identified based on their cotyledon shape and size such as tall morning glory (*Ipomoea purpurea*) or Venice mallow (*Hibiscus trionum*) (Fig. 4.4). With broadleaf species, the size of cotyledons depends on size of seed from which the plant emerges. These could be categorized into general groupings of large-, medium-, and small-seeded broadleaves. Large-seeded broadleaves often include weed species such as common cocklebur (*Xanthium strumarium*), jimsonweed (*Datura stramonium*), giant ragweed (*Ambrosia trifida*), and devil's claw (*Proboscidea louisianica*). Medium-seeded broadleaves would be velvetleaf, prickly sida (*Sida spinosa*), common ragweed (*Ambrosia artemisiifolia*), and wild sunflower (*Helianthus annuus*), while small-seeded broadleaves would include species such as pigweed species, common lambsquarters (*Chenopodium album*), and kochia.

With grass or grasslike species, the size and orientation of the first leaf can influence the ability to sense these plants from above. For example, most winter annual grasses such as downy brome (*Bromus tectorum*), wild oat (*Avena fatua*), or jointed goatgrass (*Aegilops cylindrica*) have a very upright and twisting first leaf arrangement, which creates very little green area to be sensed. Some summer annual grasses have a broad and horizontally oriented first leaf such as large crabgrass (*Digitaria sanguinalis*) or foxtail species (*Setaria* spp.) so that there is more plant material to be sensed.

Fig. 4.4 Unique cotyledon shape and first true leaf for tall morningglory (*Ipomoea purpurea)* and Venice mallow (*Hibiscus trionum*) seedlings (Photos courtesy of J.A. Dille)

Fig. 4.5 Examples of leaf arrangement on weed seedlings. (**a**) Alternate and vining leaf arrangement of wild buckwheat (*Polygonum convolvulus*). (**b**) Alternate and upright leaf arrangement of eastern blank nightshade (*Solanum ptycanthum*). (**c**) Opposite and upright leaf arrangement of giant ragweed (*Ambrosia trifida*). (**d**) Opposite and upright leaf arrangement of common sunflower (*Helianthus annuus*). (**e**) Whorled and upright leaf arrangement of kochia (*Kochia scoparia*). (**f**) Whorled and basal arrangement of marestail (*Conyza canadensis*) (Photos courtesy of J.A. Dille)

As the first true leaves begin to appear after the cotyledons, the next observation is whether the plants produce oppositely or alternately arranged leaves as well as their rate of leaf appearance. These traits influence the total leaf area that is observed and how quickly each seedling plant increases in size in the field (for example, Fig. 4.5 a–f). Most annual broadleaves have alternate leaf arrangements so those with uniquely opposite leaf arrangement include weed species in the mint family such as henbit or occur at early growth stages for common sunflower, giant ragweed, and common ragweed, members of the sunflower family (Fig. 4.5 c and d). Subsequently, first true leaf size and arrangement of opposite or alternate leaves, together with overall plant growth habit such as upright, prostrate, or vining, influence our ability to identify and "sense" small weed plants.

4 Changes in Weed Morphology

As an individual weed species develops through time, it increases in leaf number and leaf size and stem length (height) and may initiate branching. Repeatedly measuring stem length and leaf area index (leaf area/ground area, m^2/m^2) of Palmer

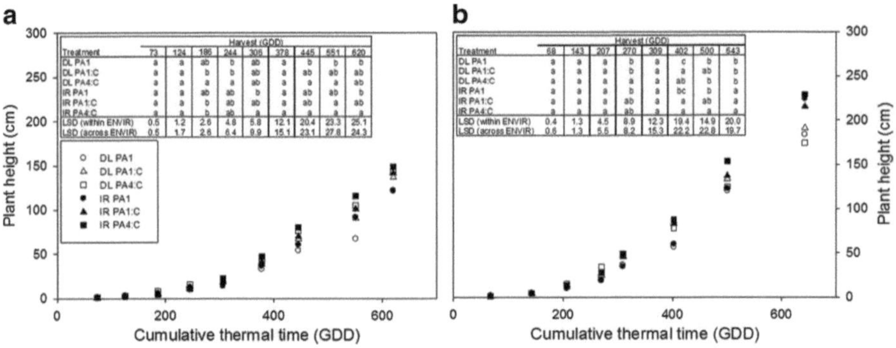

Fig. 4.6 Palmer amaranth plant height in dryland (*DL – open symbols*) and irrigated (*IR – closed symbols*) environments grown alone (PA1) at one Palmer amaranth plant m – 1 of row and with corn at one (PA1:C) or four (PA4:C) Palmer amaranth plants m – 1 of row in 2005 (**a**) and 2006 (**b**). Letter within columns by harvest date compare means using LSD (across ENVIR) (Rule 2007)

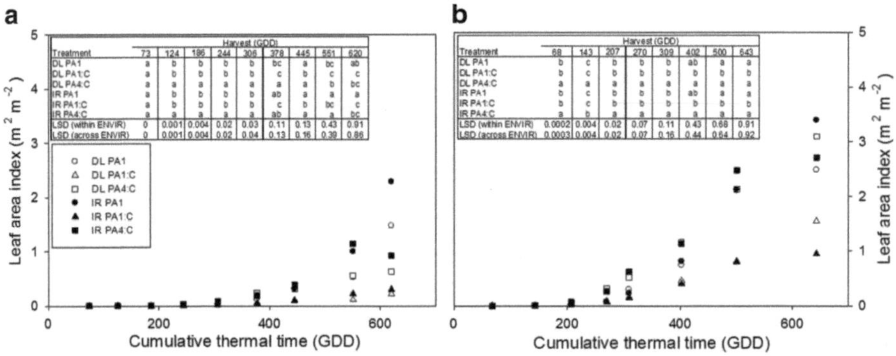

Fig. 4.7 Palmer amaranth leaf area index in dryland (*DL – open symbols*) and irrigated (*IR – closed symbols*) environments grown alone (PA1) at one Palmer amaranth plant m – 1 of row and with corn at one (PA1:C) or four (PA4:C) Palmer amaranth plants m – 1 of row in 2005 (**a**) and 2006 (**b**). Letter columns by harvest date compare means using LSD (across ENVIR) (Rule 2007)

amaranth (*Amaranthus palmeri*) over time, with or without crop competition, shows how quickly this weed species develops early in the growing season (Figs. 4.6 and 4.7). Palmer amaranth height doubled every 60 growing degree days (GDD) from emergence to 445 or 402 GDD, which is prior to corn tasseling (Rule 2007). Palmer amaranth grew and developed at a more rapid rate than common waterhemp, redroot pigweed, and tumble pigweed (*Amaranthus albus*) when grown in common conditions in Kansas (Horak and Loughin 2000). Correct identification of weed species early is critical to knowing how quickly it will increase in size, become competitive with the crop, and reach maximum size that can limit effective weed control applications (Bensch et al. 2003; Horak and Loughin 2000; Mayo et al. 1995). A generic identification of a pigweed species can be problematic if it actually

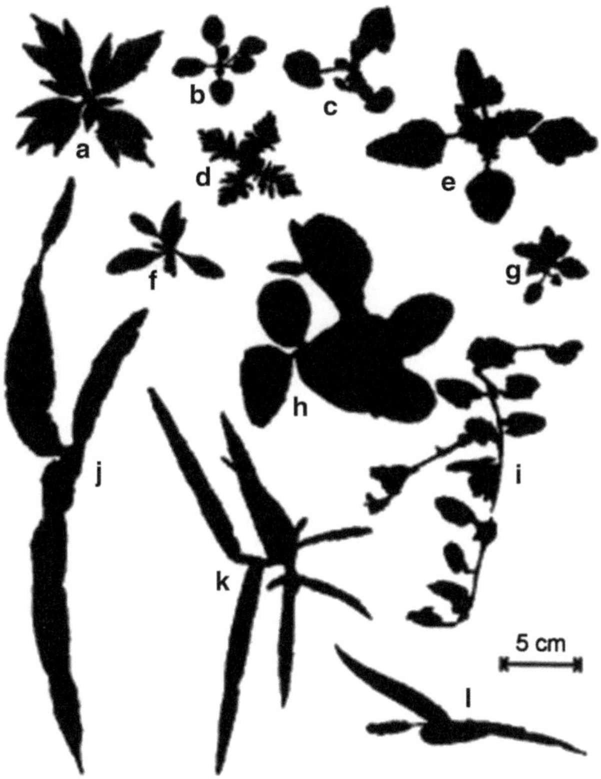

Fig. 4.8 Typical canopy shapes of various plant species 20–30 days after emergence (**a**) giant ragweed (*Ambrosia trifida*), (**b**) field pennycress (*Thlaspi arvense*), (**c**) velvetleaf (*Abutilon theophrasti*), (**d**) common ragweed (*Ambrosia artemisiifolia*), (**e**) common cocklebur (*Xanthium strumarium*), (**f**) Pennsylvania smartweed (*Polygonum pensylvanicum*), (**g**) common lambsquarters (*Chenopodium album*), (**h**) soybean (*Glycine max*), (**i**) field bindweed (*Convolvulus arvensis*), (**j**) corn (*Zea mays*), (**k**) large crabgrass (*Digitaria sanguinalis*), and (**l**) green foxtail (*Setaria viridis*) (Reprinted with permission from Woebbecke et al. 1995)

is Palmer amaranth growing at a more rapid pace than redroot pigweed, for example (Horak and Loughin 2000).

As a weed canopy develops, its geometric shape is composed of both vertical structures (stem elongation) and horizontal structures (leaf number and size, branches), which can be complex and constantly changing as the plant grows over time and with changing environmental conditions. Before the advent of most automated technologies, strategies to describe plant canopy shapes were evaluated by Woebbecke et al. (1995). Color slide photographs were taken at 2- to 4-day intervals up to 45 days after emergence of 10 common weed species as well as corn and soybean and each slide was digitized (Woebbecke et al. 1995). These digital images were used together with shape feature analysis, and the authors were able to discriminate between grasses and broadleaves at 14 and 23 days after emergence (Fig. 4.8). Canopy shape features for seedlings with few leaves are greatly affected

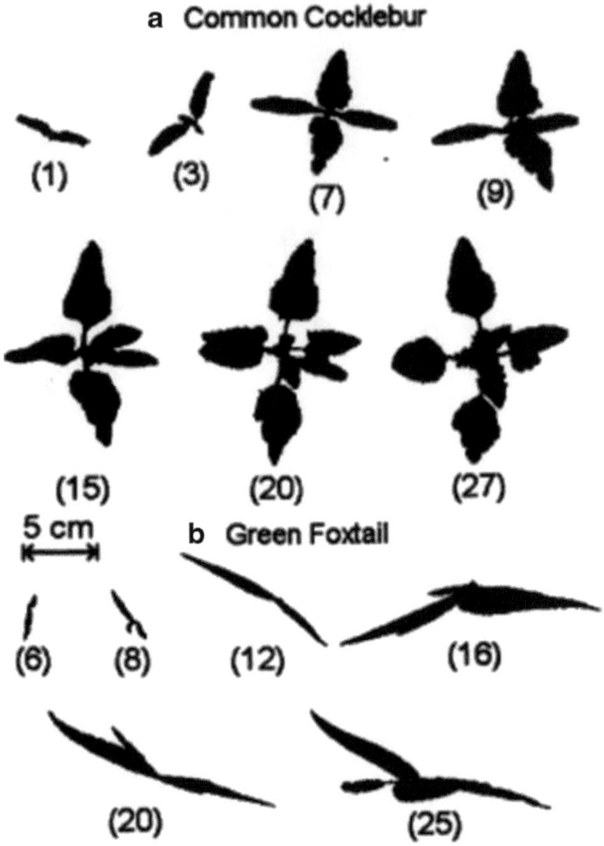

Fig. 4.9 Development of (**a**) broadleaf [common cocklebur (*Xanthium strumarium*)], and (**b**) grass [green foxtail (*Setaria viridis*)]. Each *number in parentheses* represents plant age, in days after emergence (Reprinted with permission from Woebbecke et al. 1995)

by the shape of the individual leaves. However, there is no guarantee that any particular shape feature will continue to work successfully as a plant classifier as the size of the plant increases. For example, the roundness value for monocots (calculated as $P^2/4\pi A$, where P is canopy perimeter (cm) and A is projected area (cm^2)) is close to one soon after emergence and makes it difficult to discriminate them from dicots. Some dicots emerged with cotyledons having a long slender shape (such as common cocklebur or jimsonweed), making separation from monocots difficult at this young age. The best separation (greater than 80 %) of monocots and dicots among this group of 10 weed species occurred after 14 days of age, corresponding to the full development the first true leaves in dicots (Woebbecke et al. 1995). It was determined that canopy shape features generally do not significantly change for dicots between ages of 10 and 23 days, while for monocots it was between ages 14 and 23 days (Fig. 4.9). During this window of time, seedlings of these 10 weed species are still small enough to be controlled effectively with most chemical options and occur early in the CPWC for most crops.

The rate of change in canopy shape will be influenced by the rate of leaf appearance, which is driven primarily by temperature in a linear relationship (Alm et al. 1988;

Cao and Moss 1989) and corresponds to the normal range of temperatures observed during the growing season. For example, Gramig and Stoltenberg (2007) determined the mean thermal time needed for the rate of appearance of each leaf for six common weed species was 37.2 GDD per leaf for giant ragweed, 34.4 GDD for velvetleaf, 17.3 GDD for redroot pigweed, 42.2 GDD for large crabgrass, 65.2 GDD for woolly cupgrass, and 34.2 GDD for wild-proso millet. Shrestha and Swanton (2007) determined the rate of leaf appearance for 4 other weed species to be 20–25 GDD per leaf for common lambsquarters, 50–100 GDD for barnyard grass, 14–25 GDD for redroot pigweed, and 33–50 GDD for wild mustard (*Sinapis arvensis*). The rate of leaf appearance for the weed species was not influenced by the crop planting date; environmental conditions (moisture and temperature) at crop planting have a greater influence on the initial rate of leaf appearance of weeds. At the time of scouting (sensing) for weed species, knowledge of how quickly a weed species produces individual leaves would allow for prediction of future ground area covered by knowing leaf size and density of the weed species in a given area.

5 Crop Losses from Weeds According to CPWC

Numerous studies have determined crop yield losses, reductions in crop quality, or inefficient harvesting due to effects of weed densities and time of weed emergence relative to the crop. These studies have provided critical threshold densities as to when to implement weed control practices. Next is to determine the best time frame to scout a crop and determine if the weed population or community is greater than this threshold. This time frame is described by the critical period of weed control (CPWC), defined as the time period during the life cycle of the crop when it must be kept weed-free to prevent yield loss due to weed competition (Knezevic et al. 2002; Swanton and Weise 1991). The CPWC can be experimentally derived by determining two components: (A) duration of weed interference and (B) weed-free period (Fig. 4.10). The time period when these two components overlap is known as the CPWC, and it has a distinct beginning (crop growth stage, CGS_x) and end (CGS_y) (Fig. 4.10).

The "duration of weed interference" (A) is the length of time weeds can be present with the crop from the time of crop planting before causing greater than the threshold level of yield loss and thus need to be controlled. The length of the "weed-free period" (B) is determined by how long the crop must be maintained weed-free before later emerging weeds that grow with the crop do not cause more than the threshold level of yield loss. The level of yield loss used to predict the beginning and end of the CPWC can range from 2 to 10 % (Hall et al. 1992; Refsell 2013; Van Acker et al. 1993) and could be selected based on cost of weed control and anticipated financial gain of controlling the weeds. The CPWC for corn, soybean, winter wheat, dry bean, canola, and lentils has been derived experimentally with the unique conditions that influence the beginning and end (Table 4.1). Zimdahl (1980, 2004) has also summarized the CPWC for many other agronomic and horticultural crops.

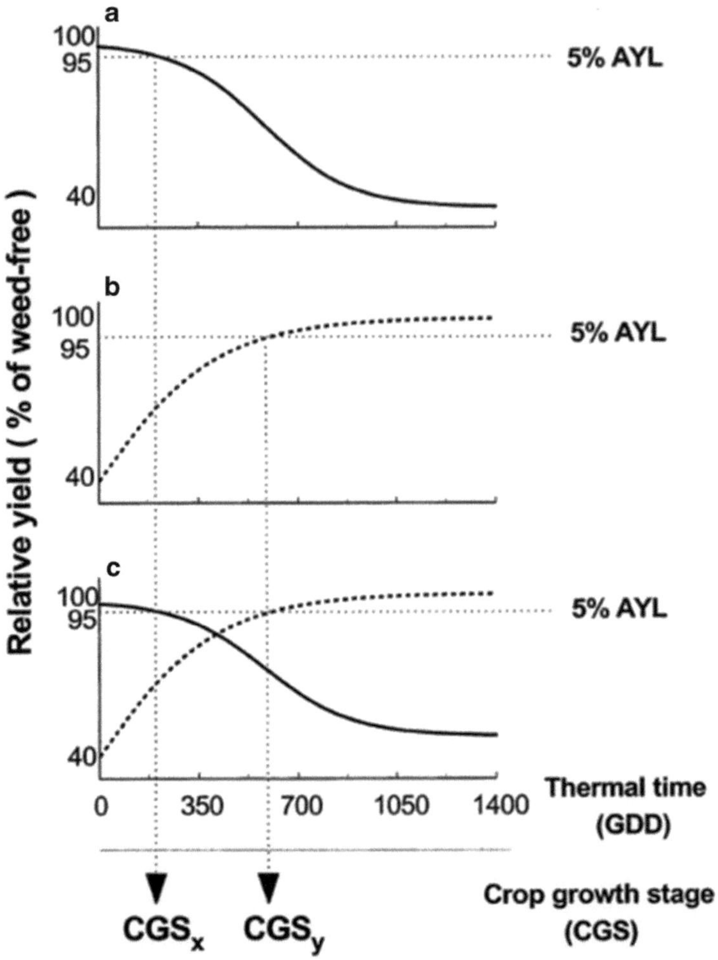

Fig. 4.10 Functional approach used for the determination of the critical period of weed control (CPWC). (**a**) The critical timing for weed removal determined from the logistic model, fit to data representing an increasing duration of weed interference. (**b**) The critical weed-free period is determined from the Gompertz model, fit to data representing an increasing duration of weed-free period. (**c**) The value of the *x* axis that corresponds to 95 % relative yield or an acceptable yield loss (*AYL*) of 5 % is determined for both curves and related to the crop growth stage (*CGS*). The CPWC is then defined as the time period between the two crop growth stages (CGS_x to CGS_y) and represents the length of weed control required to protect crop yield from more than a 5 % yield loss (Reprinted with permission from Knezevic et al. 2002)

Thus, the CPWC provides a time frame for when to assess a weed population or community. Prior to the beginning of the CPWC, field scouting (sensing) for weeds must be initiated. It is at this time when the weed species in the field need to be identified and their density and size determined (Fig. 4.11). If field scouting for weeds reveals that there is a weed population or community of concern, and at a sufficient

Table 4.1 The beginning (duration of weed interference A) and end (weed-free period B) of the critical period of weed control as determined for different agronomic crops, locations, and anticipated yield loss threshold levels

Crop	Location	Weed species[a]	Yield threshold (%)	Duration of weed interference (A)	Weed-free period (B)	Reference
Corn	Ontario	Mixed annuals	2.5	3–6 WAP[b]	7 WAP	Hall et al. (1992)
	Texas	Johnson grass	5	3 WAP	6.5 WAP	Ghosheh et al. (1996)
	Wisconsin	Wild-proso millet	5	V2 CGS	V11 CGS	Mickelson and Harvey (1999)
	Nebraska	Mixed annuals	5	4 WAP	5–7 WAP	Evans et al. (2003)
	Ontario	Mixed annuals	2.5	2–2.5 WAP	3.5–6.5 WAP	Halford et al. (2001)
Soybean	Ontario	Mixed annuals	2.5	2–5 WAP	4 WAP	Van Acker et al. (1993)
	Illinois	Common waterhemp		2–4 WA unifoliate expansion	–	Hager et al. (2002)
	Nebraska	Mixed annuals	5	V1 (1 WAP)	–	Knezevic et al. (2003)
	Ontario	Mixed annuals	2.5	2–3 WAP	5.5–6 WAP	Halford et al. (2001)
Dry bean	Minnesota	Mixed annuals	5	3 WAP	5–6 WAP	Burnside et al. (1998)
	Ontario	Mixed annuals	3	4 WAP	7.5 WAP	Woolley et al. (1993)
Winter wheat	Kansas	Downy brome	5	19–29 WAP	5 WAP	Refsell (2013)
	England	Mixed annuals	5	4 WAP	20 WAP	Welsh et al. (1999)
Spring canola	Manitoba	Mixed annuals	10	2.5–5.5 WAP	2.5–5.5 WAP	Martin et al. (2001)
Lentil	Saskatchewan	Mixed annuals	5	5-node CGS	10-node CGS	Fedoruk et al. (2011)

[a] Johnson grass (*Sorghum halepense*), wild-proso millet (*Panicum miliaceum*), common waterhemp (*Amaranthus rudis*), downy brome (*Bromus tectorum*)
[b] *WAP* weeks after planting, *CGS* crop growth stage

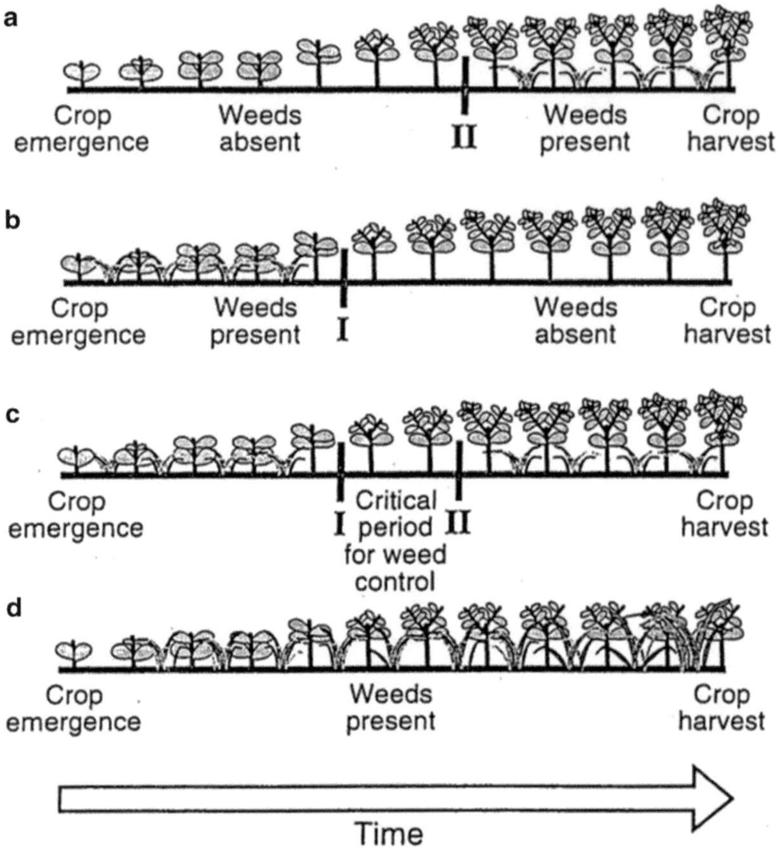

Fig. 4.11 Apparent CPWC. (**a**) If weeds are absent up to point II, crop dominance is established and yield losses do not result, even though weeds may be present subsequently. (**b**) If weeds are present for a period of time following crop emergence but are absent for the remainder of the season, yield losses do not result since, presumably, early in the season weeds are too small for competition to occur. (**c**) The combination of results from (**a**) and (**b**) leads to the critical period between points I and II, "window" of time during which weeds must be removed or suppressed to avoid crop yield loss at harvest. (**d**) Situation in which weeds are present throughout the growing season and crop yield loss results (Reprinted with permission from Radosevich et al. 2007)

density to eventually cause crop loss if left uncontrolled (greater than 2–10 % anticipated yield loss or a weed density that is greater than an economic threshold), then control measures must be implemented. Some chemical weed control practices take some time to show their effects; thus, timely removal of weeds just prior to the beginning of the CPWC is needed to ensure that no competitive weeds are present during the CPWC. Toward the end of the CPWC, late emerging weeds need to be scouted for and, based on the species identity and their density, be evaluated as to whether further control strategies need to be planned and implemented.

The start and length of the CPWC are influenced by time of weed seedling emergence relative to the crop and weed seedling density. Weeds that emerge later

Table 4.2 Mean (±SE) weed leaf numbers at the soybean (Glycine max) second trifoliate stage in relation to time of weed emergence, with data pooled over 1997, 1998, and 1999

	Weed cohort[a]		
Weed species	VE	V1	V2
Velvetleaf (*Abutilon theophrasti*)	4.6 (0.1)	–[b]	1.7 (0.1)
Common lambsquarters (*Chenopodium album*)	9.0 (0.4)	7.1 (0.2)	2.7 (0.2)
Common ragweed (*Ambrosia artemisiifolia*)	7.5 (0.2)	4.0 (0.2)	1.2 (0.1)
Common cocklebur (*Xanthium strumarium*)	5.7 (0.1)	3.7 (0.1)	2.2 (0.2)
Redroot pigweed (*Amaranthus retroflexus*)	7.4 (0.2)	3.5 (0.2)	–
Green foxtail (*Setaria viridis*)	6.8 (0.2)	3.8 (0.1)	1.3 (0.1)
Giant foxtail (*Setaria faberi*)	6.3 (0.1)	3.8 (0.2)	1.1 (0.1)
Barnyard grass (*Echinochloa crus-galli*)	6.3 (0.1)	5.2 (0.1)	2.0 (0.3)
Fall panicum (*Panicum dichotomiflorum*)	4.8 (0.1)	3.1 (0.1)	–

Reprinted with permission from Weaver (2003)

[a] The VE cohort emerged before the soybean unifoliate stage; the V1 cohort emerged between the unifoliate and the first trifoliate; the V2 cohort emerged between the first and the second trifoliate

[b] Weed species not present in this cohort

beneath a crop canopy are much less competitive than those that emerge at the same time as the crop (Dieleman et al. 1995; Liphadzi and Dille 2006). It can be difficult to determine the time a weed species emerges, and this is important with respect to knowing how competitive the weed will be in the crop and its potential impact on final yields. A scout can compare the number of leaves of a weed species and the corresponding growth stage of the crop to determine when it might have emerged (Weaver 2003). For example, in Ontario, Canada, the CPWC for soybean is between the first and third trifoliate (V1 to V3). By the second trifoliate growth stage, velvetleaf emerging with soybean had an average of 4.6 (±0.1) leaves per plant, while those that emerged just prior to V2 growth stage of soybean had only 1.7 leaves per plant (Table 4.2). Common lambsquarters had 9.0 leaves when emerging at VE, compared to 7.1 leaves when emerging prior to V1 and 2.7 leaves prior to V2. Similar information was determined for corn in Ontario (Table 4.3). Timing of weed emergence and relative weed competitiveness can also be affected by a soil-applied residual herbicide applied prior to crop planting such that weed emergence is delayed, and those weeds that emerge through the herbicide may be less competitive due to physiological impact of the herbicide (Liphadzi and Dille 2006).

It is difficult to scout a field and determine an average whole-field weed density because of the spatial distribution of weed populations and communities (Dille et al. 2002; Gerhards et al. 1997; Wiles et al. 1992). Weed populations are known to be patchy throughout a field with areas of none, low, and high densities; thus, decisions about whether there is a weed species of concern and at a high enough density to require control will be dependent on location within the field (Dieleman and Mortensen 1999; Lindquist et al. 1998). The use of automated technologies allows for information to be collected on the spatial distribution of weed species densities across a field, and a whole-field average weed density is no longer needed. Implementing weed control can be linked to spatial distribution of weed species densities observed during the CPWC at any location in a field.

Table 4.3 Mean (±SE) weed leaf numbers at the corn (*Zea mays*) six-leaf stage in relation to time of weed emergence, pooled over years unless otherwise indicated

	Weed cohort[a]			
Weed species	VE (1997/1998)	VE (1999)	V1	V2
Velvetleaf (*Abutilon theophrasti*)	4.6 (0.1)	3.1 (0.1)	3.1 (0.1)	1.4 (0.1)
Common lambsquarters (*Chenopodium album*)	8.5 (0.2)	–	3.4 (0.2)	1.7 (0.1)
Common ragweed (*Ambrosia artemisiifolia*)	7.3 (0.2)	5.6 (0.2)	–	2.1 (0.2)
Giant ragweed (*Ambrosia trifida*)	7.1 (0.2)	4.7 (0.2)	–	1.9 (0.1)
Common cocklebur (*Xanthium strumarium*)	8.1 (0.2)	4.9 (0.2)	4.1 (0.2)	3.0 (0.2)
Redroot pigweed (*Amaranthus retroflexus*)	8.6 (0.2)	4.9 (0.2)	–	2.5 (0.1)
Green foxtail (*Setaria viridis*)	6.7 (0.2)	4.1 (0.3)	–	3.1 (0.1)
Giant foxtail (*Setaria faberi*)	6.7 (0.1)	4.9 (0.2)	–	3.1 (0.1)
Barnyard grass (*Echinochloa crus-galli*)	6.7 (0.1)	4.7 (0.1)	–	3.4 (0.1)
Fall panicum (*Panicum dichotomiflorum*)	–[b]	–	4.1 (0.2)	1.4 (0.1)

Reprinted with permission from Weaver (2003)

[a]The VE cohort emerged before the corn two-leaf stage; the V1 cohort emerged at the corn two- to three-leaf stage; the V2 cohort emerged at the corn four- to five-leaf stage

[b]Weed species not present in this cohort

Based on economic analysis, the threshold weed density is often very low (1–5 plants m^{-2}), which might trigger the decision to implement weed control. The challenge is being able to sense and not miss these yield-impacting low densities. Armstrong et al. (2007) evaluated the use of early season multispectral images to detect low densities of common lambsquarters seedlings during the CPWC in corn. Aerial multispectral images (12–16 cm pixel resolution) were taken 18–19 and 32 days after planting in two different field studies. This resolution was much higher than previous studies which often used 0.5–1 m^2 pixel resolutions (Armstrong et al. 2007). However, with the technology at that time, they determined that corn and common lambsquarters could not be reliably detected and differentiated at either field site when weeds were 9 cm or less in height. Economic threshold densities (2 and 4 plants m^{-2}) of common lambsquarters could be distinguished from weed-free plots at one location when weeds were 17 cm in height, but unfortunately, this height was beyond that recommended on the herbicide label for glyphosate application. As technologies continue to develop, they must be able to detect very small weed seedlings before the beginning of the CPWC.

6 Crop Management and Weed Control Practices Influence the CPWC

Numerous crop management factors including the weed control practices that are already implemented can influence the beginning and end of the CPWC (Evans et al. 2003; Knezevic et al. 2002; Martin et al. 2001). Knowledge of these factors can be used to fine-tune the timing of weed scouting just prior to or during the CPWC and thus assist in determining if postemergence weed control practices are necessary to prevent a certain level of crop loss. Examples of factors include crop species present, time of crop planting, row spacing and crop seeding rates, fertility practices, and use of soil-applied residual herbicides that affect the beginning or end of the CPWC.

Crop species and its corresponding time of planting influence what weed species will co-occur in the field. A crop with rapid canopy development and row closure relative to the weed delays the beginning of the CPWC. Martin et al. (2001) observed that early seeding of canola resulted in the need for a somewhat longer weed-free period (delayed end of CPWC) because field operations were conducted relatively early in the emergence period of the weeds resulting in higher weed infestation levels for a longer time. As soybean row spacing narrowed from 76 to 19 cm, the beginning of the CPWC was delayed from V1 to V3 growth stage, highlighting that the soybean crop was developing a more competitive canopy against the weed population present (Knezevic et al. 2003). As soil N application increased from 0 to 120 kg N ha^{-1}, the start of the CPWC was delayed from V1 to V6 growth stage of corn in a Nebraska study (Evans et al. 2003; Knezevic et al. 2002).

In general, the duration of weed interference, or beginning of the CPWC, will be delayed if soil resources are abundant, the crop establishes quickly, and the weeds grow at a reduced rate. Whereas it will begin earlier if the soil resources are low, the crop establishes slowly, weed densities are high, and the weed species present grow at a rapid rate. In a similar fashion, the end of the weed-free period is short if the crop establishes quickly and is long if the crop establishes slowly. In general, develop a crop canopy that is more competitive against later emerging weeds.

7 Conclusions

The CPWC provides a time frame for when to observe and actively manage weed populations or communities in crop fields. With correct and timely observations, a combination of ecological, economic, and efficacy-based information and knowledge should be used to make weed management decisions. The CPWC occurs relatively early in the cropping season when weeds tend to be small, thus requiring improved ability for detection at a young age. Advanced technology is being developed to identify weeds with greater precision and accuracy (see Chap. 5). The use of the technology must account for variation in weed and crop growth and spatial variation in individual plants and weed populations. Knowing the conditions of weeds (e.g., growth stage) and when they transition through the different stages will allow for the development and application of automated technology. Further, the CPWC will help in determining the speed at which automated weed control systems must be able to move through the field and cover large hectares that are not static but dynamic biological systems.

References

Alm DM, Pike DR, Hesketh JD, Stoller EW (1988) Leaf area development in some crop and weed species. Biotronics 17:29–39

Anderson RL (1994) Characterizing weed community seedling emergence for a semiarid site in Colorado. Weed Technol 8:245–249

Anderson RL (2005) A multi-tactic approach to manage weed population dynamics in crop rotations. Agron J 97:1579–1583

Anderson RL, Nielsen DC (1996) Emergence pattern of five weeds in the Central Great Plains. Weed Technol 10:744–749

Armstrong JQ, Dirks RD, Gibson KD (2007) The use of early season multispectral images for weed detection in corn. Weed Technol 21:857–862

Bensch CN, Horak MJ, Peterson DE (2003) Interference of redroot pigweed (*Amaranthus retroflexus*), Palmer amaranth (*A. palmeri*), and common waterhemp (*A. rudis*) in soybean. Weed Sci 51:37–43

Burnside OC, Wiens MJ, Holder BJ, Weisberg S, Ristau EA, Johnson MM, Cameron JH (1998) Critical periods for weed control in dry beans (*Phaseolus vulgaris*). Weed Sci 46:301–306

Cao W, Moss DN (1989) Temperature effect on leaf emergence and phyllochron in barley. Crop Sci 29:1018–1021

Dieleman JA, Mortensen DA (1999) Characterizing the spatial pattern of *Abutilon theophrasti* seedling patches. Weed Res 39:455–467

Dieleman JA, Hamill AS, Weise SF, Swanton CJ (1995) Empirical models of pigweed (*Amaranthus* spp.) interference in soybean (*Glycine max*). Weed Sci 43:612–618

Dille JA, Milner M, Groeteke JJ, Mortensen DA, Williams MM II (2002) How good is your weed map? A comparison of spatial interpolators. Weed Sci 51:44–55

Dille JA, Stahlman PW, Geier PW, Riffel JD, Currie RS, Wilson RG, Sbatella GM, Westra P, Kniss AR, Moechnig MJ, Cole RM (2012) Kochia emergence profiles across the central Great Plains. Proc Weed Sci Soc Am Abstr 52:122

Elmore R (2012) Best corn planting dates for Iowa. Integrated Crop Management News and Iowa State University Extension. http://www.extension.iastate.edu/CropNews/2012/0327elmore2.htm. Accessed 31 May 2013

Evans S, Knezevic S, Lindquist JL, Shapiro C (2003) Nitrogen application influences the critical period for weed control in corn. Weed Sci 51:408–417

Fedoruk LK, Johnson EN, Shirtliffe SJ (2011) The critical period of weed control for lentil in Western Canada. Weed Sci 59:517–526

Gerhards R, Wyse-Pester DY, Mortensen DA, Johnson GA (1997) Characterizing spatial stability of weed populations using interpolated maps. Weed Sci 45:108–119

Ghosheh HZ, Holshouser DL, Chandler JM (1996) The critical period of johnsongrass (*Sorghum halepense*) control in field corn (*Zea mays*). Weed Sci 44:944–947

Gramig GG, Stoltenberg DE (2007) Leaf appearance base temperature and phyllochron for common grass and broadleaf weed species. Weed Technol 21:249–254

Hager AG, Wax LM, Stoller EW, Bollero GA (2002) Common waterhemp (*Amaranthus rudis*) interference in soybean. Weed Sci 50:607–610

Halford C, Hamill AS, Zhang J, Doucet C (2001) Critical period of weed control in no-till soybean (*Glycine max*) and corn (*Zea mays*). Weed Technol 15:737–744

Hall MR, Swanton CJ, Anderson GW (1992) The critical period of weed control in grain corn (*Zea mays*). Weed Sci 40:441–447

Hartzler RG, Buhler DD, Stoltenberg DE (1999) Emergence characteristics of four annual weed species. Weed Sci 47:578–584

Horak MJ, Loughin TM (2000) Growth analysis of four *Amaranthus* species. Weed Sci 48:347–355

Knezevic SZ, Evans SP, Blankenship EE, Van Acker RC, Lindquist JL (2002) Critical period of weed control: the concept and data analysis. Weed Sci 50:773–786

Knezevic SZ, Evans SP, Mainz M (2003) Row spacing influences the critical timing for weed removal in soybean (*Glycine max*). Weed Technol 17:666–673

Lindquist JL, Dieleman JA, Mortensen DA, Johnson GA, Wyse-Pester DY (1998) Economic importance of managing spatially heterogeneous weed populations. Weed Technol 12:7–13

Liphadzi KB, Dille JA (2006) Annual weed competitiveness as affected by preemergence herbicide in corn. Weed Sci 54:156–165

Martin SG, Van Acker RC, Friesen LF (2001) Critical period of weed control in spring canola. Weed Sci 49:326–333

Mayo CM, Horak MJ, Peterson DE, Boyer JE (1995) Differential control of four *Amaranthus* species by six postemergence herbicides in soybean (*Glycine max*). Weed Technol 9:141–147

Mickelson JA, Harvey RG (1999) Relating *Eriochloa villosa* emergence to interference in *Zea mays*. Weed Sci 47:571–577

Pedersen P (2013) Soybean planting date. Iowa State University Extension. http://extension.agron.iastate.edu/soybean/documents/PlantingDate.pdf. Accessed 31 May 2013

Radosevich SR, Holt JS, Ghersa CM (2007) Ecology of weeds and invasive plants, 3rd edn. Wiley, Hoboken, 454 pp

Refsell DE (2013) Integrated weed management in Kansas winter wheat. PhD dissertation, Department of Agronomy, Kansas State University, Manhattan, 88 pp. http://hdl.handle.net/2097/15742

Ross MA, Lembi CA (2007) Applied weed science, 3rd edn. Pearson Education, Inc., Upper Saddle, 561 pp

Rule DM (2007) Corn and Palmer amaranth interactions in dryland and irrigated environments. PhD dissertation, Department of Agronomy, Kansas State University, Manhattan, 138 pp. http://hdl.handle.net/2097/377

Shrestha A, Swanton CJ (2007) Parameterization of the phenological development of select annual weeds under noncropped field conditions. Weed Sci 55:446–454

Swanton CJ, Weise SF (1991) Integrated weed management: the rationale and approach. Weed Technol 5:648–656

Van Acker RC, Swanton CJ, Weise SF (1993) The critical period of weed control in soybean [*Glycine max* (L.) Merr.]. Weed Sci 41:194–200

Weaver SE (2003) Correlations among relative crop and weed growth stages. Weed Sci 51:163–170

Welsh JP, Bulson HAJ, Stopes CE, Froud-Williams RJ, Murdoch AJ (1999) The critical weed-free period in organically-grown winter wheat. Ann Appl Biol 134:315–320

Wiles LJ, Oliver GW, York AC, Gold HJ, Wilkerson GG (1992) Spatial distribution of broadleaf weeds in North Carolina soybean (*Glycine max*) fields. Weed Sci 40:554–557

Woebbecke DM, Meyer GE, VonBargen K, Mortensen DA (1995) Shape features for identifying young weeds using image analysis. Trans ASAE 38:271–281

Woolley BL, Michaels TE, Hall MR, Swanton CJ (1993) The critical period of weed control in white bean (*Phaseolus vulgaris*). Weed Sci 41:180–184

Zimdahl RL (1980) Weed-crop competition: a review. International Plant Protection Center, Oregon State University, Corvallis, 196 pp

Zimdahl RL (2004) Weed-crop competition: a review, 2nd edn. Blackwell Pub. Professional, Ames, 220 pp

Chapter 5
The Biological Engineer: Sensing the Difference Between Crops and Weeds

David C. Slaughter

Abstract This chapter describes the current state of the art in technology and methodology being used to develop sensors for automated weed control in cropping systems. The development of a reliable universal weed vs. crop plant sensor that works well in a wide range of crops and cropping systems is a formidable task. The discussion in this chapter highlights the significant progress that has been made in developing new, more robust, automatic sensing systems that can differentiate crop plants from weeds. Case studies documenting high levels of success in trials conducted outdoors in the natural, largely uncontrolled environment of an agricultural cropping system are presented. A discussion of the strengths and current challenges of the more successful weed and crop sensing techniques is provided. In many cases the methodology has utilized site- or condition-specific a priori knowledge to make the sensors smarter in a local context. This chapter highlights the advantages and compromises made in using these techniques. The chapter concludes with a discussion of the remaining engineering challenges to the development of a comprehensive, multifaceted fusion of several methods for sensing the differences between crops and weeds across the entire crop production cycle, and how the rapid development of advanced sensing and machine learning technologies will facilitate new plant recognition architectures and systems to achieve the level of machine recognition of weeds needed for automated weed control in cropping systems.

D.C. Slaughter (✉)
Biological and Agricultural Engineering, University of California,
2030 Bainer Hall, One Shields Avenue, Davis, CA 95616, USA
e-mail: dcslaughter@ucdavis.edu

S.L. Young and F.J. Pierce (eds.), *Automation: The Future of Weed Control in Cropping Systems*, DOI 10.1007/978-94-007-7512-1_5,

1 Introduction

Several recent overviews of the state of the art in weed sensing technologies have been conducted (e.g., Cope et al. 2012; Weis and Sökefeld 2010; Christensen et al. 2009; Slaughter et al. 2008a; Brown and Noble 2005). A key theme outlined in these reviews is that one of the major issues continuing to limit widespread implementation of automatic weed control systems in agricultural crops is the lack of a uniformly robust, automatic weed sensing method. The works by Weis and Sökefeld (2010) and Slaughter et al. (2008a) provide good overviews of the breadth of techniques that have been developed for sensing the differences between crop plants and weeds. This chapter is not intended to serve as a comprehensive review of the many methodologies that have been investigated, but rather, this chapter will focus more narrowly on those techniques that have been tested in the natural outdoor environment of an agricultural cropping system and have shown considerable promise toward more robust performance in weed detection.

In order to provide the technical details on the processes involved with a particular methodology for sensing differences between crop plant and weeds, a case study approach will be taken in this chapter, focusing on the work of a select subset of published works that illustrate the technique particularly well and in more detail, rather than giving a broad overview of all the published work on a topic. Inherent in this case study approach is that some works will not be discussed. This is not intended as an assessment on the merit of those omitted works. Readers interested in a more complete review of the literature are referred to the more general overviews mentioned in the paragraph above.

Automatically sensing weeds growing in an agricultural field prior to planting, before the crop has developed to a stage where it is readily detected, or in a fallow field, generally involves sensing live plants against a background of bare soil or against a background of soil and nonliving plant residue. This case is considerably less complex than sensing weeds growing in among a vegetative background. For the plant vs. non-plant background case, a large number of various types of fairly simple vegetation indices have been developed, and fairly mature solutions to this task are available. For interested readers, Scotford and Miller (2005) conducted a review of the more common vegetative indices used to detect live plants which are typically ratios of broadband reflectance measurements in the visible and/or near-infrared region. Mature, commercial sensors are available to detect weeds against a non-plant background (e.g., WeedSeeker, NTech Industries 2012). The focus of this chapter is on the more challenging task of sensing weeds growing in among a vegetative background, for which the task of sensing live plants against a soil background is often an initial step and will be discussed later in this chapter.

A limited number of mechanical weed sensors have been developed, such as the electromechanical plant probes by Garrett (1966) or Cox and McLean (1969) that detect plants by touch, sensing an electrically resistive path through the plant to earth ground below 10 MΩ. Such sensors can be used, for example, to selectively sense weeds that are taller than the crop (Diprose and Benson 1984). However, these types of sensors typically have several constraints limiting their use. The primary limitation of mechanical (or electromechanical) weed sensors is the need to individually touch all plants, an operation which is difficult to perform at high travel speeds without risk of mechanical damage to the crop.

By far, the most commonly utilized technologies for sensing the differences between crop plants and weeds are based upon the electromagnetic radiation properties of plants. Sensors based upon electromagnetic radiation properties are typically capable of both high-speed and noncontact measurements as well as both proximal and remote modes of operation. In addition to their electromagnetic radiation properties, the spatial context of the information in a general sense, at both the macro- and microscales, is frequently incorporated into the sensing process to improve the system performance. This contextual knowledge spans properties such as the texture of individual leaves and other plant structures, the plant morphology (size and shape of plants), and their spatial relationship to other objects or structures in the immediate surroundings.

With the appropriate optical system design (e.g., wide angle vs. telephoto lenses), a great deal of flexibility can be achieved in the range of distances permitted between the sensor and the plant when sensing a plant's optical properties. While the term remote sensing can be used to describe any noncontact sensing method, in lay terms it is generally reserved for the cases where the sensor is mounted on an airplane or satellite.

Methods of weed detection have traditionally been divided into proximal measurements taken from ground-based platforms or vehicles and remote measurements taken from airborne vehicles or satellites. When optical reflectance-based weed detection methods are used, these two detection modes share many fundamental weed detection techniques, such as image processing or spectroscopy. A review of the literature will show that, historically, aerial or satellite remote-sensing methods have been limited to lower spatial resolutions at the soil surface, often called the ground sample distance (GSD) of a remotely sensed image. However, rapid development of higher and higher spatial resolution consumer-grade color imaging sensors (e.g., ~40 megapixel image sensors by Nokia 2012 and Nikon 2012) or industrial-grade 4-band color/infrared imaging sensors (e.g., ~60 megapixel image sensor by Leica 2012) has, with the appropriate lenses, greatly improved the GSD in state-of-the-art aerial images. For example, Booth and Cox (2008, 2009) have demonstrated the feasibility of generating 1 mm GSD resolution color aerial images taken from low altitude (100 m above the soil level taken from a manned sport aircraft) using an 11 megapixel consumer-grade camera equipped with an image-stabilized telephoto lens. In a study of aerial images taken using a 4-band color/infrared camera from an altitude of ~300 m with a GSD of ~3 cm, Duniway et al. (2012) observed that a minimum of three pixels per object were required for object classification and that objects smaller than 10 cm in diameter were very difficult for human experts to classify from 3 cm GSD aerial images. The availability of low-cost unmanned aerial vehicles (UAVs, Fig. 5.1), such as the remotely controlled, miniature helicopter (e.g., Ishihama et al. 2012) that can be easily transported to a field boundary and

Fig. 5.1 Photograph of a miniature UAV helicopter equipped with a consumer grade digital color camera and GPS navigation system for low-altitude aerial imaging (Mikrokopter 2012. Reprinted with permission from http://www.mikrokopter.de)

flown at low altitude across the field using an autonomous GPS waypoint navigation system, suggests that the technology barriers to very high-resolution aerial imagery may be overcome to a large degree in the near future.

At the present, aerial-based methods for detecting weeds in vegetative agricultural fields have been successfully demonstrated for detecting weed patches, where the minimum detectable patch size is related to the GSD of the aerial imaging system deployed (Lamb and Brown 2001). For example, Lamb et al. (1999) have demonstrated detection of weeds in a seedling-stage triticale crop where the lowest population of weeds detected was 17 plants m^{-2} using aerial imagery with a GSD of 0.5 m. While high-resolution imaging sensors and low-cost UAVs may overcome some of the technological barriers to advancing the use of aerial imaging for detecting weeds at the individual plant scale, some practical limits on image quality, such as shadows from adjacent plants off-nadir in photos taken at times away from solar noon or atmospheric interference related to weather and altitude, are inherent in the method. Because of the limited amount of aerial or satellite imaging research on sensing the difference between weeds and crops at the individual plant spatial resolution scale, particularly for seedlings, and due to the general similarities in the applications of image-processing or spectroscopy techniques between aerial and ground-based sensing, the remainder of this chapter will focus primarily on ground-based methods of weed detection. Readers interested in more information specific to aerial- or satellite-based remote-sensing methods for weed detection are referred to review articles by Lamb and Brown (2001) and Brown and Noble (2005).

2 Individual Leaf Shape Sensors

A considerable body of knowledge has been amassed on developing machine vision methods of extracting and quantifying the biological morphology of plants for the purpose of species identification. When individual plant leaves are carefully displayed, where the leaf is constrained to a 2D plane that is perpendicular to the optical axis of the camera, the leaf morphology and vein structure can be observed (Fig. 5.2a). In this situation, where there is very high contrast in either monochrome or color between the background and the object, 2D image-processing techniques can be readily applied to segment the leaf from the background and to extract the leaf morphology. Readers are referred to recent books on computer vision and image-processing methods (e.g., Burger and Burge 2008; Davies 2012; Sonka et al. 2007)

Fig. 5.2 (**a**) Photograph reprinted with permission from Cope et al. (2012) showing examples of different leaf shapes. (**b**) Application of a Sobel edge extraction for leaf margins and veins. (**c**) Pseudo color representation of (**b**) to highlight the leaf texture and veins

Table 5.1 Examples of leaf shape and leaf texture features for the leaf images in Fig. 5.2

Row	Leaf[a]	Circularity[b]	Aspect ratio	Edge variance	Median edge strength
1	1	0.18	1.39	567	227
1	2	0.76	1.19	492	222
1	3	0.22	1.52	628	237
1	4	0.74	1.21	816	239
1	5	0.64	1.05	660	238
2	1	0.48	1.90	313	240
2	2	0.44	1.78	257	240
2	3	0.13	1.10	265	239
2	4	0.50	1.84	112	242
2	5	0.55	1.27	1,166	238
3	1	0.45	2.97	152	237
3	2	0.55	1.34	117	243
3	3	0.61	1.08	2,252	235
3	4	0.74	1.20	661	238
3	5	0.65	1.55	375	223

[a]Indicates the column, from 1 on the left to 5 on the right in Fig. 5.2a
[b]Circularity as defined by http://imagej.nih.gov/ij/docs/guide/146.html, where Circularity $=4\pi *$Area/Perimeter^2

for detailed descriptions of classical methods for image enhancement and object segmentation under these conditions. In this ideal case, where the object pose, the background, and illumination are controlled, leaf segmentation, using either color or intensity, is robust and edge detection (both the leaf margins and veins) is easily obtained, providing both the leaf boundary and interior texture (Fig. 5.2b, c). A number of methods have been developed to mathematically characterize the leaf shape using object shape properties or leaf boundary curvature (Table 5.1). More successful leaf shape recognition methods utilize normalization techniques that make the shape feature calculations less sensitive to planar orientation, starting point on the leaf boundary trace, and scale.

Cope et al. (2012) published a recent review of a number of image-processing approaches to recognize the species from individual leaf images (e.g., Fig. 5.2). Recent works by Lee and Chen (2006) and Du et al. (2007) give examples of 2D object shape features useful in leaf shape characterization and demonstrate the performance of a shape feature classifier on sets of 60 and 20 plant species, respectively, using 2D images of individual leaves. A number of studies have been published on the use of leaf boundary curvature from individual 2D leaf images for plant species recognition (e.g., Franz et al. 1991; Chi et al. 2002). As an example of a recent work in an agricultural context, Neto et al. (2006) used elliptic Fourier descriptors of a closed contour of the leaflet edge calculated from the 2D chain code of the 2D image to classify soybean, sunflower, and two weed species. A species classification rate of 89.4 % was achieved on images taken in the third week after germination. They commented that species classification was difficult at the cotyledon stage due to similarities in the cotyledon leaf shapes for these species. In a much larger study, Hearn (2009) described a method using leaf shape for plant species recognition in a study of 51 plant species from the Sonoran Desert and 100 species from a lowland tropical rainforest. Using a combination of Fourier and Procrustes analyses of the leaf boundary pixels, Hearn found that a minimum of 10 leaves per species, 100 leaf margin points (i.e., spatial resolution), and 10 Fourier harmonics were required to accurately characterize the 2D leaf shape of a species. This method was 94 % accurate in recognizing Sonoran Desert species but dropped to an accuracy of 74 % in recognizing leaves from the rainforest. Hearn observed that the 2D leaf shapes of desert species were more distinct than the "notoriously elliptic" leaf shapes of many rainforest species. Hearn also tabulated the species recognition accuracy for a number of recent studies using other machine vision methods for 2D leaf analysis. The species recognition rates varied from 59 to 96 % depending upon the number and type of species involved and the number of leaves examined.

For automated weed control, however, 2D images of isolated leaves that capture the leaf shape from an ideal pose (i.e., as shown in Fig. 5.2) are not readily available in situ. Two common factors leading to degraded machine vision performance in shape recognition are the visual occlusion of leaf shape due to overlapped leaves and the nonoptimal display of leaf shape for automated machine vision methods when the leaves are still attached to the plant and photographed in situ. Franz et al. (1991) conducted one of the few studies that has characterized the impact of nonideal leaf pose (i.e., where the leaf is tilted away from the ideal image plane) and the loss of leaf shape information due to leaf occlusion (i.e., where part of the leaf is hidden behind another object, typically another part of the plant) when the species template database comes from the ideal pose (i.e., Fig. 5.2a). In this study, Franz et al. found that leaf species was correctly recognized from a portion of the leaf edge curvature when leaf occlusion was less than 20 % (i.e., at least 80 % of the leaf shape was visible). With regard to leaf pose, they found that species recognition was not impaired as long as the leaf plane was positioned within a 30-degree angle from the plane perpendicular to the camera's optical axis. Lee (1998) conducted a similar study on the impact of cotyledon leaf orientation in tomato when distinguishing tomato seedlings from grass using a single top-view image. He found that as the view changed from an ideal orthographic projection, similar to Fig. 5.2a, to a perspective view of the leaf shape caused by a nonideal leaf pose, the ability to distinguish tomato cotyledons from grass failed when the angle of the tomato cotyledons was greater than 62° from the horizontal.

3 Individual Plant Shape Sensors

Machine vision methods that are more robust to the natural variation in leaf pose for 2D images taken in situ have been attempted. These methods fall under the general machine vision, pose-invariant, object recognition topic which includes the "face in a crowd" body of work on face recognition in computer vision dealing with biological

Fig. 5.3 Illustrations demonstrating the concept of pose invariant seedling recognition using deformable pattern templates. (**a**) shows a possible leaf template (in *red*) overlaid on a plant leaf, where the *green arrows* indicate the template deformations required to make a better shape match. (**b**) shows the deformed template from (**a**) after adjustment. An image of a fat-hen (*Chenopodium album*) is overlaid (in *red*) with initial pattern templates for dead-nettle (*Lamium* spp.) and fat-hen (*Chenopodium album*) in (**c**) and (**e**), respectively. The appearances of the templates from (**c**) and (**e**) are shown after deformation in (**d**) and (**f**), respectively (Reprinted with permission from Søgaard 2005)

systems observed in situ. One method for plant recognition that has demonstrated some success in automatically adapting the pattern recognition process to the specific pose in each image is based upon initial work by Kass et al. (1988) using the concept of a dynamic or "active" view of a contour. Several published works on the application of this active shape recognition concept to leaf edge boundaries have been conducted (e.g., Manh et al. 2001; Søgaard 2005; Persson and Åstrand 2008).

To illustrate the concepts of deformable species pattern templates and pose-invariant leaf shape recognition, the method and illustrations from Søgaard (2005) will be used. In his work, Søgaard started with RGB color images of 19 Danish weed seedlings at the cotyledon and 1st true leaf stage acquired in situ. The Excessive Green color index (2*Green-Red-Blue) was used to identify green objects (presumably plants) in each image, and overall plant size was used to select the green subjects for the species recognition step. The species classification step involved superimposing a candidate species' model template (shown as the red boundary curve) over the actual leaf (shown as the light gray object) (Fig. 5.3a). An iterative process was then conducted where the template was translated, rotated, rescaled, and deformed to better match the actual leaf edge contour (Fig. 5.3b).

Two examples from Søgaard's work are provided, showing (in red) the overlay of the initial pattern templates for dead nettle (*Lamium* spp.) and fat hen (*Chenopodium album*) weeds on top of a fat hen seedling (Fig. 5.3c, e), respectively, and the final version of the deformed templates (Fig. 5.3d, f), respectively. Species classification decisions were then based upon a weighted combination of the amount of template deformation required to obtain an optimal fit and the final level of match between the deformed template and the unknown plant.

The performance of the deformable species pattern template method has been investigated for plant species recognition by several researchers. Mahn et al. (2001) applied the technique to leaf recognition of green foxtail seedlings, where some of the leaves were partially occluded by adjacent foxtail seedlings. In this work on a single species (*Setaria viridis*), 84 % of ~600 leaves were successfully matched by the deformed templates. Søgaard (2005) applied the technique to whole weed seedlings, viewed in isolation (i.e., one plant per image, as shown in Fig. 5.3c–f) using a validation set of images of plants from three weed species (shepherd's purse, *Capsella bursa-pastoris* (L.) Medik.; scentless mayweed, *Tripleurospermum Schultz-Bip.*; and charlock, *S. arvensis* L.). Correct species classification rates of 77 %, 65 %, and 93 %, respectively, for the three species were achieved. The method shows promise as a means of implementing species template matching on seedlings viewed in situ. However, performance was adversely affected by off-target placement of the initial overlay of the template on the object and by missing leaves (Persson and Åstrand 2008).

3.1 Plant Recognition: Using Leaf Veins

Nam et al. (2008) developed a single leaf vein pattern matching system for use with a web-based automatic species identification system. Their system characterized the vein texture using the number of vein intersections and endpoints and their internodal distances. When trained on 1,032 leaf images taken from the botany guidebook *Illustrated Flora of Korea*, they were able to automatically classify leaf vein patterns into pinnate, parallel, and palmate venation classes with 90 %, 94 %, and 86 % recall rates from the database, respectively.

An example of a more advanced use of leaf vein information in the automatic recognition of partially occluded cauliflower seedlings has been developed by Soille (2000a, b) (Fig. 5.4). Soille's approach is a fusion of several image-processing techniques and provides an excellent example showing how multifaceted knowledge of a crop and the growing environment may be exploited to achieve a successful technique. The method begins with the color segmentation of plants and soil. Soille approached this step using nonparametric clustering techniques (Fig. 5.4b); however, in this case a naive Bayesian classifier assuming a bivariate normal distribution for plants in the red/green color space is equally effective. Cauliflower has whitish leaf veins and the bottom of the leaf is very light green in color. Being low in color saturation, the pixels for the veins, exposed leaf bottom, and heavily shadowed

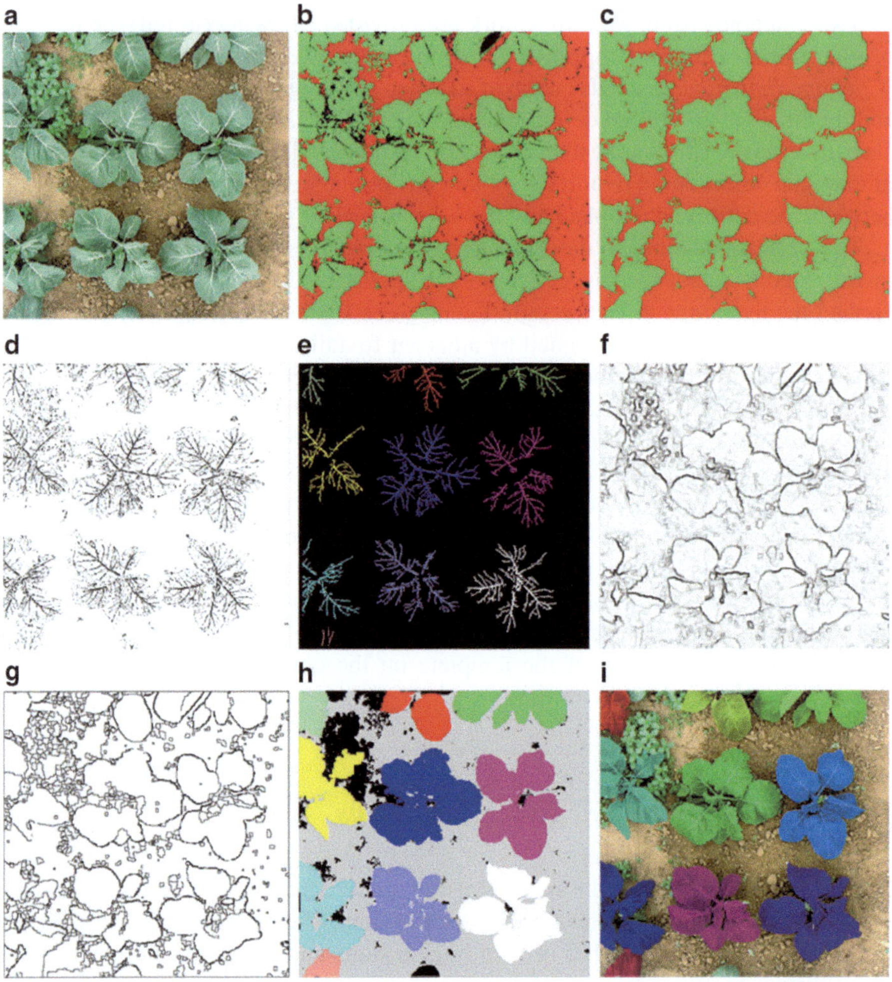

Fig. 5.4 Example image sequence reprinted with permission from Soille (2000b) showing the automatic segmentation of partially occluded cauliflower (*Brassica oleracea* var. *botrytis*) seedlings from weeds and soil. (**a**) Original color image. (**b**) Plant classification using the Watershed feature space, (**c**) Improved plant classification after application of a majority rule for black objects in (**b**). (**d**) Thresholded White Top-Hat image. (**e**) Hierarchical vein clustering. (**f**) Morphological gradient of the intensity image. (**g**) Watersheds of the segmentation function. (**h**) Labeled vegetation catchment basins. (**i**) Final pseudo color segmentation of cauliflower plants

pixels were not well segmented using color, and Soille improved the segmentation by applying a spatial neighbor majority rule to create the improved color segmentation (Fig. 5.4c).

The leaf veins in this example were then segmented (Fig. 5.4d) by applying the mathematical morphology White Top-Hat operator (Soille 1999) to the raw intensity image (Fig. 5.4a). A key to the success of this step is the exploitation of knowledge about the scene, which is then manifested in the choice of the size of the structuring element for the White Top-Hat operator that was selected to be slightly larger than the size of the largest cauliflower leaf vein. Top-Hat operators are a filtering technique where a morphological opening or closing step is used to remove the image structure of interest (in this case the veins) and then to recover them by subtracting the resulting image from the original image. White Top-Hat operators are used to extract bright objects, while the Black Top-Hat will extract dark objects. They are useful in naturally illuminated scenes where gradual illumination gradients may be present and can also filter noise from the image. Soille performed additional morphological filtering steps followed by an intensity-weighted hierarchical clustering step to label the veins belonging to each plant (Fig. 5.4e). This method was quite effective at extracting the veins in this context because the crop plants were small enough that their foliage did not significantly overlap one another and were, for the most part, not hidden from view.

Once the plant-by-plant vein clusters were extracted, they could be used to help identify the boundaries of partially occluded plants. This technique was quite effective in this case because the crop plant foliage was predominately located above the weed plant foliage, making the veins of crop plants visible even when the cauliflower leaf boundaries were not. To segment the cauliflower plants from the weeds, first an intensity gradient "edge" filter was applied followed by edge threshold filtering and then the watershed object segmentation function (Fig. 5.4f and g). The previously extracted crop plant veins (Fig. 5.4e) were then used as spatial markers to show which watershed catchment basins (Fig. 5.4g) should be recombined to eliminate the oversegmentation of the crop plant foliage by the watershed algorithm (Fig. 5.4h). Soille then applied the label mask (Fig. 5.4h) to create the pseudo-colored version of the original image (Fig. 5.4i). The method was very effective at recognizing the heavily occluded cauliflower leaf (colored cyan) shown in the upper left corner of Fig. 5.4i).

3.2 Plant Recognition: Using Texture

A number of researchers have explored the use of visual texture to detect weeds. Visual texture can be extracted at different scales, ranging from the texture within an individual leaf (e.g., Fig. 5.2c) to the whole plant level. Many studies have demonstrated good performance in identifying weed species using visual texture of whole plants. A number of researchers have investigated visual texture based upon the spatial dependence method, which determines the conditional probability of occurrence of a particular color or gray level at a vector location from the starting point (Haralick et al. 1973). The name "co-occurrence matrix" is used for the resulting matrix of conditional probabilities created for all combinations of each pair of co-occurring grayscale or color values at the defined spatial offset and direction. In the majority of the published studies investigating the use of texture for weed detection, the images contained a monoculture (i.e., a single species per image) and classification rates of 90 % or more were frequently obtained (Shearer and Holmes 1990; Meyer et al. 1998; Burks et al. 2002; Tang et al. 2003; Ishak et al. 2009).

In the special case, where the crop is a monocot (flowering plants where the embryo has a single cotyledon) and the weeds are dicots (flowering plants where the embryo has two cotyledons), a number of successful texture-based techniques have been published that show good performance in handling partially occluded scenes. Dating from the early work by Ahmad et al. (1991) on finding dicot weeds in a grass lawn, a number of automatic methods have been developed based upon the simple premise stated by Watchareeruetai et al. (2006) "that the grass area should contain a lot of edges while the weed area is smoother than the grass area." This basic premise then provides the a priori context knowledge about the relative size of the monocot crop leaves to the size of the dicot leaves. An approach, similar to that used by Soille (2000a), where the size of the structuring element is carefully selected based upon this a priori knowledge, can be used to extract weeds in partially occluded scenes using a simple sequence of mathematical morphology steps.

The use of color, edge detection, and mathematical morphology can detect partially occluded weeds (*Oxalis corniculata*) in a lawn (Fig. 5.5). As observed by Gebhardt et al. (2006), the use of color helps to distinguish live plant material from soil or other background objects that have a level of homogeneity in visual texture similar to the broadleaf weeds. The first step in this process is to extract the gradient image (Fig. 5.5b) from the raw intensity image (Fig. 5.5a). A color segmentation mask is then used to remove any objects in the image that are not similar in color to the broadleaf weeds (Fig. 5.5c). The remaining grass foliage was then filtered by thresholding the edge gradient to select homogeneous texture regions and then applying neighborhood-based, median filtering using a circularly shaped structuring element whose size was selected based upon the size of the *Oxalis* weed leaf size. The outlines of the detected weed objects were then overlaid on the original image (Fig. 5.5d).

Gebhardt and Kühbauch (2007) used textural homogeneity as a visual feature to distinguish three dicot weed species, broad-leaved dock (*Rumex obtusifolius*), dandelion (*Taraxacum officinale*), and broadleaf plantain (*Plantago major*), from ryegrass (*Lolium perenne* L.) in grassland images. While the foliage of the dicot species was intermingled with the grass and partial cross-species occlusion occurred, the view of the *Rumex obtusifolius* plants was mostly unoccluded, due to their larger size and positioning above the grass foliage. A greater degree of cross-species occlusion occurred with the other two dicots. In this work, localized texture homogeneity was calculated within a 5×5 pixel structural element using the intensity gradient and standard deviation images. Regions of homogeneous texture were extracted by thresholding the homogeneity image, and color was used to distinguish soil from plants. They then evaluated a set of 17 machine vision features including object shape (area, perimeter, eccentricity, roundness, and shape factor), texture (local mean and standard deviations of the gradient, homogeneity, and standard deviation images), and color (both mean and standard deviation values). The local homogeneity texture feature had the best weed discrimination power followed by the remaining texture and color features, with most of the misclassifications occurring between the dicot species. They concluded that the object shape features were ineffective at providing additional classification value for distinguishing the dicot weeds

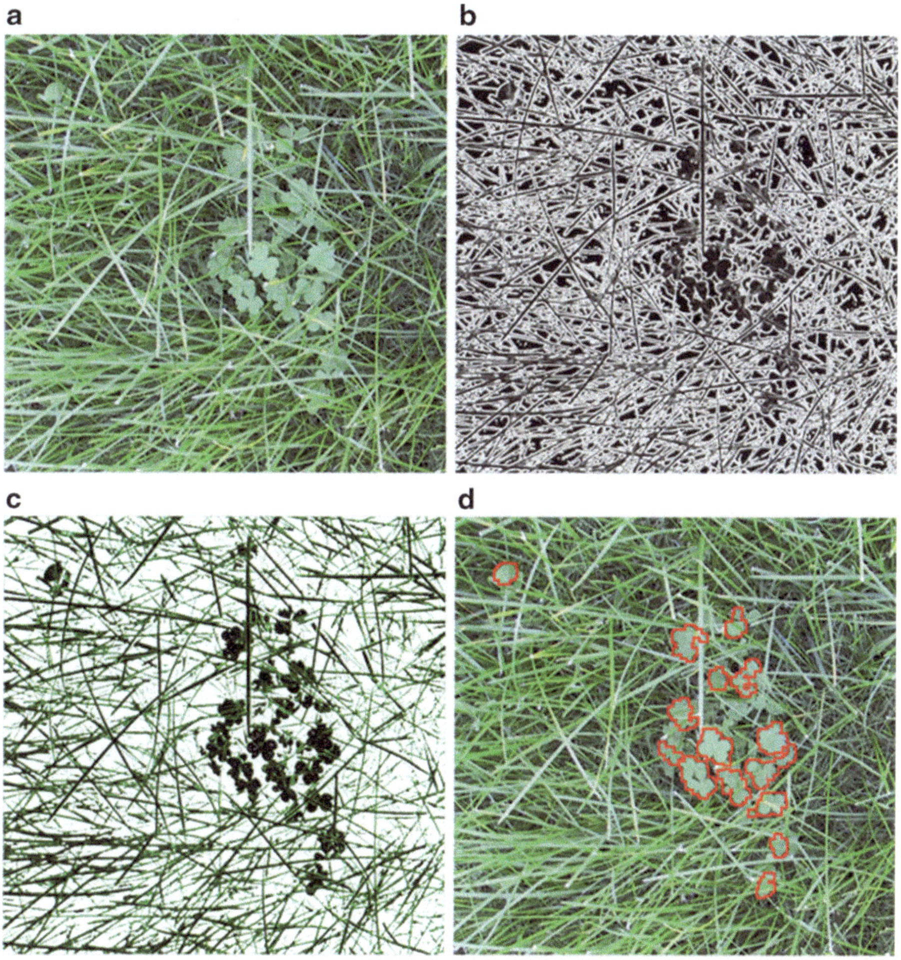

Fig. 5.5 Example showing the use of visual texture to detect broadleaf weeds (*Oxalis corniculata*) in grass in the presence of visual occlusion. (**a**) Original color image. (**b**) Sobel edge image of the intensity image. (**c**) Color segmentation of be to remove non-green objects. (**d**) Detected broadleaf weeds outlined in *red* based upon mathematical morphology filtering of (c)

from grass in this context due to partial leaf occlusion and the associated inability to accurately characterize the leaf shape. A classifier using only the 12 features related to object texture and color was able to successfully identify 93 % of the broad-leaved dock plants. Only 1.4–4.7 % and 3.9–9.6 % of broad-leaved dock plants were misclassified as broadleaf plantain or dandelion, respectively, depending upon growth stage. Similar levels of dandelion and broadleaf plantain were misclassified as *Rumex obtusifolius*.

Another image texture-based method to distinguish partially occluded broad-leaved dock plants in grassland images was developed by van Evert et al. (2009). As with the other texture-based methods, a localized texture feature was used for weed recognition, which in this case was based upon the 2D FFT power spectrum calculated within 8×8 pixel tiled regions of the image. Object area and power spectrum thresholding were used to segment the broad-leaved dock weeds from the grass. Broad-leaved dock weeds were correctly identified in 23 of 28 grassland images (82 % detection rate). The five false-negatives were reported to be caused when images were taken of broad-leaved dock that had recently been grazed by animals on the dairy farm where the images were collected, so that only a few, small leaves of broad-leaved dock remained.

3.3 Plant Recognition: Using Spectroscopy

The optical reflectance characteristics of plants, particularly in the visible and near-infrared regions (400–2,500 nm), provide an attractive means of distinguishing crop plants from weeds in situ. Without the need to observe the leaf boundary, as required in shape-based methods, this type of method is robust to partial occlusion. More general reviews of this topic have been done by Brown and Noble (2005) and Scotford and Miller (2005). This section will focus on the more advanced applications of the technique to field-grown plants.

Lamb and Brown (2001) have stated that "spectral reflectance characteristics in general do not vary significantly among plant species during vegetative growth stages (Price 1994) making discrimination between weeds and crop plants impossible on that basis alone." However, the referenced work by Price (1994) was based upon a univariate analysis of the root mean square differences in reflectance spectra averaged across the entire spectrum from ~500 to ~2,300 nm which observed that "spectra from one species may match very closely spectra from another species, presenting the possibility of faulty identification." Advanced multivariate techniques, such as canonical discriminant analysis, were not attempted, and the proportion of cases of accurate vs. faulty species identification was not documented. Differences in plant morphology and differences in species-related plant response to growing conditions can be manifested in changes in biochemical composition, cell size, and developmental structure, all of which can create species-related characteristics in the spectral reflectance of whole plants.

While the hope by some that spectral reflectance would provide a species-unique, time-, and growing condition-invariant spectral fingerprint for all plant species may be unrealistic (Zwiggelaar 1998), a growing body of scientific research over the past decade has demonstrated the feasibility of distinguishing crop and weed species on a site-specific basis using advanced multivariate spectral reflectance techniques. For example, Slaughter et al. (2004) conducted a spectral reflectance study of field-grown tomato and nightshade leaves, which are from the same taxonomic family (*Solanaceae*). Visual examination of the mean reflectance spectra for tomato and black nightshade from this work appears quite similar (Fig. 5.6), so it is easy to see why a casual assessment of the spectral differences might conclude that

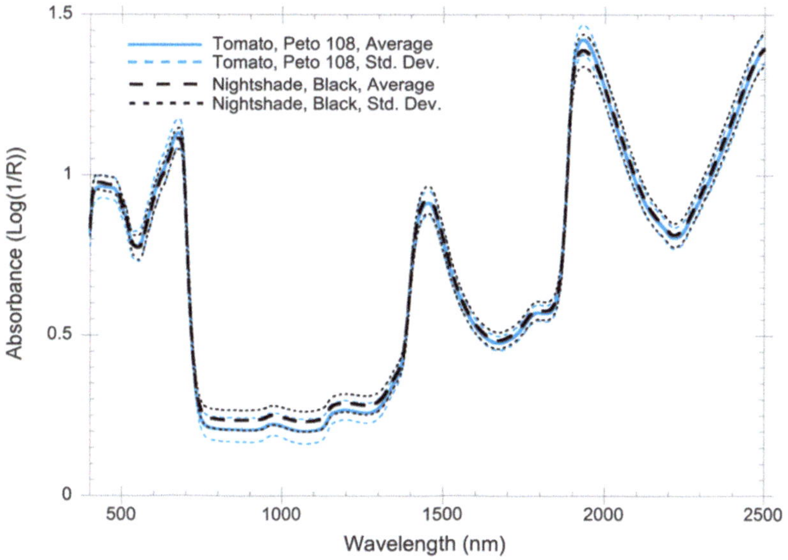

Fig. 5.6 Mean absorbance spectra for tomato (*solid curve*) and black nightshade (*dashed curve*) leaves measured using diffuse reflectance geometry. The absorbance standard deviation at each wavelength is shown by the thin dotted curves (Reprinted with permission from Slaughter et al. 2004)

significant differences were not present. Slaughter et al., however, demonstrated that a broadband RGB color classifier based upon the visible portion of these spectra could, on average, accurately classify by species 76 % of plants in one-out cross-validation tests. Further, narrowband multivariate classifiers based upon canonical discriminant analysis using near-infrared reflectance could accurately classify plants by species with accuracies of 98 % or higher in one-out cross-validation tests. In a study by Galvão et al. (2005) of four sugarcane cultivars grown on farms in Southeastern Brazil where the mean reflectance spectra of the cultivars were also visually similar, application of a multivariate, canonical discriminant classifier was able to correctly classify 90 % of the pixels in three cultivars and 80 % of the pixels in the fourth. The classifier was trained on visible and near-infrared wavebands sensitive to differences in chlorophyll, water, and lignin-cellulose contents as well as sunlight penetration into the canopy related to canopy architecture.

For automatic weed control, the leaf reflectance must be measured in situ, an environment much less controlled than the laboratory method used by Slaughter et al. (2004). In their review, Lamb and Brown (2001) noted that several research studies have found that differences in plant species growth habit, growth rates, vigor, presence of tomentose, reproductive stage, or senescence may enhance the spectral differences between weeds and crop plants and aid in their species classification. A number of researchers have demonstrated the infield use of ground-based, visible, and near-infrared hyperspectral line-imaging systems for in situ plant species recognition.

Fig. 5.7 Ground-based hyperspectral imaging system setup for controlled illumination of crop and weed seedlings. (**a**) Photograph showing the tractor-mounted metal illumination control chamber in the center of the picture. (**b**) Interior view of the enclosure showing the hyperspectral camera and the illumination system

Feyaerts and van Gool (2001), Vrindts et al. (2002), and Okamoto et al. (2007) have all demonstrated the successful use of a spectral reflectance-based classifier collected using a push-broom-type hyperspectral imager in sugar beet fields. Classification accuracies ranged from 80 to 95 % in the two-class, crop vs. weed type of classifier used in the studies by Feyaerts and van Gool and Vrindts et al. Okamoto et al. developed an individual species classifier and reported classifier validation accuracies of 81.3 % for sugar beets and 74.7–89.3 % classification accuracies for the four weed species. Another common feature of these three systems was that all used natural sunlight as the illumination source. Although a spectral reference standard was utilized to normalize the reflectance spectra for temporal changes in illumination and instrument sensitivity, Vrindts et al. observed that the validation performance was substantially degraded when there was a change in the natural illumination from direct sunlight to a more diffuse illumination under a partly cloudy sky.

To overcome the problems associated with variability in natural illumination, Slaughter et al. (2008b) developed a tractor-mounted hyperspectral imaging system that utilized a metal enclosure to block the natural sunlight and a controlled illumination system to provide spatially uniform and time-invariant illumination of the scene (Fig. 5.7). This controlled illumination, hyperspectral imaging system was operated outdoors in lettuce fields containing both crisphead and leaf lettuce plants as well as three weed species at the second true leaf stage of growth. A multivariate classifier, based upon visible and near-infrared reflectance, was able to correctly classify the crisphead, leaf lettuce, and three weed species at 83 %, 87.9 %, and 75.9–86.3 % accuracy levels, respectively, in 1 %-out cross-validation tests. When the classifier was modified to perform a two-class, crop vs. weed classification, the cross-validation accuracies improved to 89.4 %, 94 %, and 83.6–92.5 %, respectively.

The effect of environmental stressors, such as temperature, moisture, or solar radiation levels, on the optical properties of plants has been studied by a number of researchers (Guyot 1990). In tomato, for example, Nightingale (1933) observed that anthocyanin (which absorbs light in the visible region) accumulation in the plant foliage was inversely related to growing temperature. Henry et al. (2004) compared spectral reflectance classifier performance of three soil moisture-specific classifiers for soybean from common cocklebur (*Xanthium strumarium* L.) and sicklepod (*Cassia obtusifolia* L.). The models achieved classification accuracies of 86 %, 91 %, and 91 % for the control (no stress), moderate, and high moisture stress trained classifiers, respectively. In a similar study in tomato, Zhang (2011) investigated the impact of drought stress on species classification performance for tomato, black nightshade, and pigweed using multivariate classifiers based upon visible and near-infrared reflectance measurements. Zhang observed species-dependent changes in the spectral reflectance at the canopy level due to moisture stress, where accumulation of pigments, such as anthocyanin, degradation of chlorophyll, or changes in leaf size or plant morphology varied by species. In both of these cases, the authors observed that the plant species involved were more distinct when drought stress occurred, resulting in some improvement in species identification rates under stress. In studies examining either variations in diurnal temperature, solar radiation levels, or moisture conditions, Zhang and Slaughter (2011a, b) and Zhang (2011) found that a global calibration procedure, where the training set included reflectance spectra from plants grown across the full range of environmental conditions, provided fairly robust and stable species classification performance, with species recognition rates of ~90 % when validated across a wide range of growing temperatures and moisture stress conditions. Peñuelas et al. (1993) observed that changes in spectral reflectance due to plant drought stress were more evident at the canopy level than at the leaf level, supporting comments by Lamb and Brown (2001) that utilization of canopy-level reflectance measurements may be beneficial in species classification.

In one of the few multi-season, infield studies using a visible and near-infrared spectral reflectance-based classifier for plant species recognition, Zhang et al. (2012) developed a ground-level, visible, and near-infrared hyperspectral weed mapping system for black nightshade and pigweed at the seedling stage in tomato fields. Separate naïve Bayesian classifiers developed for each season individually produced good in-season performance with cross-validation species pixel classification rates above 92 %. Cross-season validation of these single season-based classifiers showed dramatically degraded performance due to changes, primarily in the NIR, in the plant reflectance spectra. A machine learning approach was developed where three artificial intelligence (AI) season "expert" classifiers were used in a multi-season, multi-classifier approach. The species classification decisions of the

Fig. 5.8 Image reprinted with permission from Southall et al. (2002) showing a pseudocolor crop (*green plants*) and weed (*red plants*) mapping result based upon crop planting pattern recognition of transplanted cauliflower

three seasonal AI experts agreed 77.4 % of the time. A "mixture-of-experts" seasonal species classifier approach was then applied which automatically adjusted the weight placed upon the knowledge learned in the previous seasons by the three seasonal AI experts. This approach achieved an across season species recognition rate of 95.8 %, which was slightly higher than a global calibration approach that had multi-season species recognition accuracies ranging from 90 to 92.7 %.

3.4 Plant Recognition: Using Spatial Contexts

Most high-valued annual crops are precision planted in rows, using a fixed row-to-row spacing as well as a fixed plant-to-plant spacing along the row. Crop emergence and subsequent crop size are fairly uniform in a well-managed commercial field where water and nutrients are not limiting. In some crop species, or with transplanted crops, there may be a consistent crop vs. weed species difference in morphological stature (i.e., upright vs. sprawling growth habit) or in plant age that can be exploited in distinguishing crop plants from weeds. In these cases, a localized spatial context exists in two or three dimensions where crop planting pattern or plant stature or both can be utilized for weed detection and mapping.

In a 2D top view or downward perspective view of the field (e.g., Fig. 5.8), it has been suggested that the "as planted" crop plant locations follow a bivariate normal spatial distribution (Onyango and Marchant 2003). In the time period between planting and canopy closure, the regular, visual pattern of crop plant objects can form the basis of a crop vs. weed classifier that is more broadly applicable to a wide range of cropping situations than many other machine vision-based crop vs. weed feature sets. Preliminary development of a crop/weed classifier using the planting pattern concept was conducted by a group of researchers at the former Silsoe Research Institute in the UK as part of the Autonomous Horticultural Vehicle program (Southall et al. 2002). In many of the original Silsoe studies, transplanted cauliflower was used as the target crop, and the field of view typically included three or more crop rows with three or more cauliflower plants per row being visible (e.g., Fig. 5.8). A soil vs. vegetation segmentation step, possibly followed by noise reduction using mathematical morphology or area thresholding, was used to create an image of vegetative objects. The object locations were then compared to the expected planting pattern and subsequently classified into weeds or crop plants based upon their position. The Silsoe group has published a number of papers describing several approaches to the basic method (e.g., Tillett et al. 2001, 2008; Southall et al. 2002; Onyango and Marchant 2003) and a commercial product called the Robocrop has been developed based upon the concept (Tillett and Hague Technology Ltd. 2005). Recognition rates for transplanted cauliflower plants ranged from 82 to 96 %, and weed recognition rates ranged from 68 to 95 %. The authors report that crop plant misclassifications were mostly caused by failing to cluster all leaves into a single crop plant object, and some unclustered crop leaves were then classified as weeds. For weeds, the majority of errors occurred when weeds were close to the crop plant and clustered in with the crop object.

Åstrand and Baerveldt (2005) have applied the 2D planting pattern context-based weed detection method to direct-seeded sugar beet fields at both the first true leaf and cotyledon stages, with weed densities of 50 weeds/m^2 and 400 weeds/m^2, respectively. Further, the crop emergence rate was about 70 % in these studies. Sugar beet detection rates of 93.9 % and 61.1 % were obtained in field images, respectively, at the first true leaf and cotyledon stages. The crop recognition results at the larger plant stage were comparable to those obtained by the Silsoe group on transplanted cauliflower. However, at high weed densities and with the crop at the cotyledon stage, the ability to accurately identify the planting pattern diminished. Weed recognition rates of 64.3 % and 83.7 % were obtained in these images with weed densities of 50 weeds/m^2 and 400 weeds/m^2, respectively. In general, the 2D planting pattern context method is tolerant to a modest amount of occlusion since most implementations do not require a complete view of the leaf or plant margin. Like the method of Soille (2000a), the spatial context method can be optimized if the relative sizes of the crop and weed plants and the planting pattern are known a priori. However, good performance requires the establishment of a uniform crop stand in order to consistently recognize the planting pattern in the image. The 2D spatial context method is similar to GPS-based crop mapping strategies (e.g., Ehsani et al. 2004; Nørremark et al. 2007; Sun et al. 2010) in terms of both the ability to be readily adapted to a wide range of crops with minimal crop-specific knowledge and suitability to mechanical, blocking-type, intra-row weeding tools (e.g., Pérez-Ruiz et al. 2012). The method is, however, restricted to early growth stages in the period prior to canopy closure and to low or modest weed densities. As weed densities and crop plant sizes increase, additional steps would need to be added, such as the watershed blob splitting and leaf vein clustering steps in Soille (2000a) to handle problems with occlusion and merging of crop and weed clusters.

More recently there have been a few examples using spatial context in three dimensions to classify crop plants and weeds. While ultimately systems that can reconstruct the complete 3D morphology of plants may be developed and allow for more sophisticated 3D plant sensors, current studies have focused primarily on situations where there are distinct species differences in plant size or height. Andújar et al. (2011) used the reflected echo in a time-of-flight ultrasound method to classify broad-leaved weeds and grasses by height with classification accuracies ranging from 77.8 to 98.5 % depending upon plant species and age. Mixtures of broad-leaved weeds and grass were not well recognized due to the fairly low (0.3 m) spatial resolution of the measurement in the field of view. Haff et al. (2011) successfully demonstrated an x-ray-based system to sense the location of the main stem of transplanted tomatoes, where both plant height and stem diameter were greater in crop plants than the weeds due to the more advanced age of the crop in a transplanted field.

Piron et al. (2009, 2011) developed a 3D stereo machine vision system to determine pixel-based plant height using the coded, structured illumination method (Fig. 5.9). The height of each plant pixel above the soil surface formed the basis of a weed vs. crop classifier. An evaluation study was conducted in an outdoor plot of direct-seeded carrot that also contained several weed species including *Sonchus asper* (L.), *Chenopodium*, *Achenocloa*, *Cirsium*, *M. perennis*, *Brassica*, and *Matricaria maritime*. By restricting the analysis to a narrowband spatial region in the image about the seedline and within a time period between 21 and 30 days after planting, an automatic method could be employed to both detect the carrot seedline and determine the range of plant heights within that region corresponding to carrots vs. weeds. This step was important to successful unsupervised determination of the carrot plant height both because there were weeds present that were both shorter and taller than the crop and because the optimal height classification thresholds changed with plant growth over time. Classification accuracies ranged from 75 to 84 % for carrots and 84–88 % for weeds using the 3D height-based classifier. Piron et al. (2009) also explored the sensor fusion of plant height and broadband (50–80 nm bandwidths) multispectral features to determine if adding multispectral information could improve classification performance. In this study, they observed only a slight improvement, from 82 to 84 %, in overall classification accuracy when the broadband multispectral information was added to the classifier. While no attempt was made, beyond pixel height, to extract higher-level leaf shape or leaf texture information from the 3D depth images (Fig. 5.9), the future potential of the 3D stereo imaging technique for extracting the 3D shape of upper leaf surfaces from a vegetative layer below is very apparent.

Fig. 5.9 Pseudocolor 3D depth image adapted from Piron et al. (2009). Higher valued pixels (*red* and *orange colors*) are closer to the camera than lower valued pixels (*green* and *blue colors*). The 3D method illustrates the potential value in 3D image methods for dealing with overlapped foliage, which typically causes leaf shape recognition problems in a 2D top view of the scene

4 Conclusions

Over the past decade, significant progress has been made to develop new, more robust, automatic sensing systems that can differentiate crop plants from weeds. There have been several published accounts documenting high levels of success in trials conducted outdoors in the natural, largely uncontrolled environment of an agricultural cropping system. This chapter has highlighted several of the more successful techniques, showing how in many cases the authors utilized site- or condition-specific a priori knowledge to make the sensors smarter in a local context. Often these performance gains come at the expense of reduced ease of generalization of the methodology to other sites or conditions.

The majority of the research effort in developing methods for sensing the difference between crops and weeds has been directed at the early crop growth stages. There are two motivations for this focus on the early growth stage. First, the detrimental impact on the crop due to weeds is typically greatest at the growth stage during the first month of crop development (see Chap. 4). Second, when the plants are small, there is less foliage occlusion and thus a higher chance of observing the plants as individual objects making the task of automatic recognition simpler.

As the plants grow and mature, there is typically more intermingling of the foliage of adjacent plants and a high degree of foliage occlusion. Of the methods described in this chapter, shape-based recognition methods are the ones most severely challenged by occlusion which is then reflected in the focus of many studies on the early growth stages before adjacent plants touch. The methods that have demonstrated the greatest success in sensing the difference between crops and weeds at later growth stages are those that do not rely on observing the complete boundary of each object such as visual texture (e.g., textural differences between narrow-leafed monocots and broad-leafed dicots), or spectral reflectance (particularly when the development rate or stage of development, e.g., reproductive or senescence, of the crop is different from the weeds), or when there is a different plant density and collective plant pattern (particularly when weeds grow in fairly large and dense patches). It remains an engineering challenge to develop a comprehensive, multifaceted fusion of several methods for sensing the differences between crops and weeds across the entire crop production cycle.

While the development of a universal weed vs. crop plant sensor that is both very accurate and works well in a wide range of crops and cropping systems is indeed a formidable task, the rapid development of advanced sensing and machine learning technologies will make reaching this goal more realistic. The extremely rapid development of digital cameras and graphical processing units (GPUs) for the entertainment and consumer products industries will continue to bring higher-resolution and higher-quality imaging sensors and massively parallel processing power into the marketplace at a relatively low cost. Combined with teraflop computational power currently available in a thumb-sized processor (e.g., Intel's Xeon Phi coprocessor), it becomes more and more realistic to envision a mobile platform with a sensor system capable of creating high spectral resolution, high spatial resolution, and 3D reconstruction of a plant in real time. Future use of massively parallel GPUs could also facilitate plant recognition architectures that employ a multi-expert, contextual, cropping-system-specific, and site-specific machine learning system to achieve weed recognition performance across multiple crops and cropping systems currently achieved in the specific crop and weed pairings highlighted in this chapter.

References

Ahmad U, Kondo N, Arima S, Monta M, Mohri K (1991) Weed detection in lawn field using machine vision. J JSAM 61(2):61–69

Andújar D, Excolà A, Dorado J, Fernández-Quintanilla C (2011) Weed discrimination using ultrasonic sensors. Weed Res 51:543–547

Åstrand B, Baerveldt A-J (2005) Plant recognition and localization using context information and individual plant features. Paper IV In: Vision based perception for mechatronic weed control. Åstrand, B. PhD thesis. Chalmers University of Technology. Göteborg, Sweden. fou.sjv.se/fou/download.lasso?id=Fil-000788. Last accessed 24 Sept 2012

Booth DT, Cox SE (2008) Image-based monitoring to measure ecological change. Front Ecol Environ 6:185–190

Booth DT, Cox SE (2009) Dual-camera, high-resolution aerial assessment of pipeline revegetation. Environ Monit Assess 158:23–33

Brown RB, Noble SD (2005) Site-specific weed management: sensing requirements – what do we need to see? Weed Sci 53:252–258

Burger W, Burge MJ (2008) Digital image processing. An algorithmic introduction using Java. Springer, New York. ISBN 978-1-84628-379-6

Burks TF, Shearer SA, Green JD, Heath JR (2002) Influence of weed maturity levels on species classification using machine vision. Weed Sci 50(6):802–811

Chi YT, Chien CF, Lin TT (2002) Leaf shape modeling and analysis using geometric descriptors derived from Bezier curves. Trans Am Soc Agric Eng 46(1):175–185

Christensen S, Søgaard HT, Kudsk P, Nørremark M, Lund I, Nadimi ES, Jørgensen R (2009) Site-specific weed control technologies. Weed Res 49:233–241

Cope JS, Corney D, Clark JY, Remagnino P, Wilkin P (2012) Plant species identification using digital morphometrics: a review. Expert Syst Appl 39:7562–7573

Cox SWR, McLean KA (1969) Electro-chemical thinning of sugar beet. J Agric Eng Res 14(4):332–343

Davies ER (2012) Computer and machine vision: theory, algorithms, practicalities, 4th edn. Elsevier, Waltham, 871 p. ISBN 9780123869081

Diprose MF, Benson FA (1984) Electrical methods of killing plants. J Agric Eng Res 30:197–209

Du JX, Wang XF, Zhang GJ (2007) Leaf shape based plant species recognition. Appl Math Comput 185:883–893

Duniway MC, Karl JW, Schrader S, Baquera N, Herrick JE (2012) Rangeland and pasture monitoring: an approach to interpretation of high-resolution imagery focused on observer calibration for repeatability. Environ Monit Assess 184:3789–3804

Ehsani MR, Upadhyaya SK, Mattson ML (2004) Seed location mapping using RTK-GPS. Trans Am Soc Agric Eng 47:909–914

Feyaerts F, van Gool L (2001) Multi-spectral vision system for weed detection. Pattern Recognit Lett 22:667–674

Franz E, Gebhardt MR, Unklesbay KB (1991) Shape description of completely visible and partially occluded leaves for identifying plants in digital images. Trans Am Soc Agric Eng 34(2):673–681

Galvão LS, Formaggio AR, Tisot DA (2005) Discrimination of sugarcane varieties in Southeastern Brazil with EO-1 Hyperion data. Remote Sens Environ 94:523–534

Garrett RE (1966) Development of a synchronous thinner. J Am Soc Sugar Beet Technol 14(3):206–213

Gebhardt S, Kühbauch W (2007) A new algorithm for automatic *Rumex obtusifolius* detection in digital images using colour and texture features and the influence of image resolution. Precis Agric 8:1–13

Gebhardt S, Schellberg J, Lock R, Kühbauch W (2006) Identification of broad-leaved dock (*Rumex obtusifolius* L.) on grassland by means of digital image processing. Precis Agric 7:165–178

Guyot G (1990) Optical properties of vegetation canopies. In: Steven MD, Clark JA (eds) Applications of remote sensing in agriculture. Butterworths, London, pp 19–44

Haff RP, Slaughter DC, Jackson ES (2011) X-ray based stem detection in an automatic tomato weeding system. Appl Eng Agric 27(5):803–810

Tillett and Hague Technology Ltd. (2005) Autonomous crop treatment vehicle. http://www.thtechnology.co.uk/Past%20Projects.html. Last accessed 24 Sept 2012

Haralick RM, Shanmuga K, Dinstein I (1973) Textural features for image classification. IEEE Trans Syst Man Cybern SMC3(6):610–621

Hearn DJ (2009) Shape analysis for the automated identification of plants from images of leaves. Taxon 58:934–954

Henry WB, Shaw DR, Reddy KR, Bruce LM, Tamhankar HD (2004) Spectral reflectance curves to distinguish soybean from common cocklebur (*Xanthium strumarium*) and sicklepod (*Cassia obtusifolia*) grown with varying soil moisture. Weed Sci 52(5):788–796

Ishak AJ, Hussain A, Mustafa MM (2009) Weed image classification using Gabor wavelet and gradient field distribution. Comput Electron Agric 66:53–61

Ishihama F, Watabe Y, Oguma H (2012) Validation of a high-resolution, remotely operated aerial remote-sensing system for the identification of herbaceous plant species. Appl Veg Sci 15:383–389

Kass M, Witkin A, Terzopoulos D (1988) Snakes: active contour models. Int J Comput Vis 1:321–331

Lamb DW, Brown RB (2001) Remote-sensing and mapping of weeds in crops. J Agric Eng Res 78:117–125

Lamb DW, Weedon MM, Rew LJ (1999) Evaluating the accuracy of mapping weeds in seedling crops using airborne digital imaging. Avena spp. in seedling triticale. Weed Res 39(6):481–492

Lee WS (1998) Robotic weed control system for tomatoes. PhD dissertation. University of California, Davis, 287 p

Lee CL, Chen SY (2006) Classification of leaf images. Int J Imaging Syst Technol 16:15–23

Leica (2012) RCD30 Series, 60MP single camera head delivers co-registered, multispectral RGBN imagery. http://www.leica-geosystems.com/en/Leica-RCD30_86987.htm. Last accessed 24 Sept 2012

Manh AG, Rabatel G, Assemat L, Aldon MJ (2001) Weed leaf image segmentation by deformable templates. J Agric Eng Res 80(2):139–146

Meyer GE, Mehta T, Kocher F, Mortensen DA, Samal A (1998) Textural imaging and discriminant analysis for distinguishing weeds for spot spraying. Trans Am Soc Agric Eng 41(4):1189–1197

Mikrokopter (2012) XL Hexakopter (MK Hexa XL). http://gallery.mikrokopter.de/main.php?g2_view=dynamicalbum.PopularAlbum&g2_itemId=63419&g2_imageViewsIndex=0. Last accessed 24 Sept 2012

Nam Y, Hwang E, Kim D (2008) A similarity-based leaf image retrieval scheme: joining shape and venation features. Comput Vis Image Underst 110:245–259

Neto J, Meyer G, Jones D, Samal A (2006) Plant species identification using elliptic Fourier leaf shape analysis. Comput Electron Agric 50:121–134

Nightingale GT (1933) Effects of temperature on metabolism in tomato. Bot Gaz 95:35–58

Nikon (2012) Nikon D800 36.3MP FX-format CMOS sensor. http://www.nikonusa.com/en/Nikon-Products/Product/Digital-SLR-Cameras/25480/D800.html. Last accessed 24 Sept 2012

Nokia (2012) Nokia 808 PureView 41.0 megapixel camera sensor. http://www.nokia.com/us-en/products/phone/808/specifications/. Last accessed 24 Sept 2012

Nørremark M, Søgaard HT, Griepentrog HW, Nielsen H (2007) Instrumentation and method for high accuracy georeferencing of sugar beet plants. Comput Electron Agric 56:130–146

NTech (2012) WeedSeeker automatic spot spraying system. http://www.ntechindustries.com/weedseeker-home.html. Last accessed 24 Sept 2012

Okamoto H, Murata T, Kataoka T, Hata SI (2007) Plant classification for weed detection using hyperspectral imaging with wavelet analysis. Weed Biol Manag 7:31–37

Onyango CM, Marchant JA (2003) Segmentation of row crop plants from weeds using colour and morphology. Comput Electron Agric 39:141–155

Peñuelas J, Filella I, Biel C, Serrano L, Save R (1993) The reflectance at the 950–970 nm region as an indicator of plant water status. Int J Remote Sens 14:1887–1905

Pérez-Ruiz M, Slaughter DC, Gliever CJ, Upadhyaya SK (2012) Automatic GPS-based intra-row weed knife control system for transplanted row crops. Comput Electron Agric 80:41–49. doi:10.1016/j.compag.2011.10.006

Persson M, Åstrand B (2008) Classification of crops and weeds extracted by active shape models. Biosyst Eng 100:484–497

Piron A, Leemans V, Lebeau F, Destain MF (2009) Improving in-row weed detection in multispectral stereoscopic images. Comput Electron Agric 69:73–79

Piron A, van der Heijden F, Destain MF (2011) Weed detection in 3D images. Precis Agric 12:607–622

Price JC (1994) How unique are spectral signatures? Remote Sens Environ 49:181–186

Scotford IM, Miller PCH (2005) Applications of spectral reflectance techniques in Northern European cereal production: a review. Biosyst Eng 90(3):235–250

Shearer SA, Holmes RG (1990) Plant identification using color co-occurrence matrices. Trans Am Soc Agric Eng 33(6):1037–1044

Slaughter DC, Lanini WT, Giles DK (2004) Discriminating weeds from processing tomato plants using visible and near-infrared spectroscopy. Trans Am Soc Agric Eng 47(6):1907–1911

Slaughter DC, Giles DK, Downey D (2008a) Autonomous robotic weed control systems: a review. Comput Electron Agric 61:63–78

Slaughter DC, Giles DK, Fennimore SA, Smith RF (2008b) Multispectral machine vision identification of lettuce and weed seedlings for automated weed control. Weed Technol 22(2):378–384

Søgaard HT (2005) Weed classification by active shape models. Biosyst Eng 91(3):271–281

Soille P (1999) Morphological image analysis: principles and applications. Springer, Berlin, 316 p

Soille P (2000a) Morphological image analysis applied to crop field mapping. Image Vis Comput 18:1025–1032

Soille P (2000b) Morphological image analysis applied to crop field mapping: electronic appendix. http://mdigest.jrc.ec.europa.eu/soille/ivc2000/elecapp.html. Last accessed 24 Sept 2012

Sonka M, Hlavac V, Boyle R (2007) Image processing, analysis, and machine vision, 3rd edn. Thomson Engineering, Toronto, 972 pp. ISBN 13: 978–0495082521

Southall B, Hague T, Marchant JA, Buxton BF (2002) An autonomous crop treatment robot: Part I. A Kalman filter model for localization and crop/weed classification. Int J Robot Res 21(1):61–74

Sun H, Slaughter DC, Pérez-Ruiz M, Gliever C, Upadhyaya SK, Smith RF (2010) RTK GPS mapping of transplanted row crops. Comput Electron Agric 71:32–37

Tang L, Tian L, Steward BL (2003) Classification of broadleaf and grass weeds using Gabor wavelets and an artificial neural network. Trans Am Soc Agric Eng 46(4):1247–1254

Tillett ND, Hague T, Miles SJ (2001) A field assessment of a potential method for weed and crop mapping on the basis of crop planting geometry. Comput Electron Agric 32:229–246

Tillett ND, Hague T, Grundy AC, Dedousis AP (2008) Mechanical within-row weed control for transplanted crops using computer vision. Biosyst Eng 99:171–178

van Evert FK, Polder G, van der Heijden GWAM, Kempenaar C, Lotz LAP (2009) Real-time vision-based detection of *Rumex obtusifolius* in grassland. Weed Res 49:164–174

Vrindts E, De Baerdemeaeker J, Ramon H (2002) Weed detection using canopy reflection. Precis Agric 3(1):63–80

Watchareeruetai U, Takeuchi Y, Matsumoto T, Kudo H, Ohnishi N (2006) Computer vision based methods for detecting weeds in lawns. Mach Vis Appl 17:287–296

Weis M, Sökefeld M (2010) Chapter 8: Detection and identification of weeds. In: Oerke E-C et al (eds) Precision crop protection – the challenge and use of heterogeneity. Springer, Dordrecht, pp 119–134. doi:10.1007/978-90-481-9277-9_8

Zhang Y (2011) Hyperspectral vision-based machine learning for robust plant recognition in autonomous weed control. PhD dissertation. University of California, Davis

Zhang Y, Slaughter DC (2011a) Influence of solar irradiance on hyperspectral imaging-based plant recognition for autonomous weed control. Biosyst Eng 110(3):330–339. doi:10.1016/j.biosystemseng.2011.09.006

Zhang Y, Slaughter DC (2011b) Hyperspectral species mapping for automatic weed control in tomato under thermal environmental stress. Comput Electron Agric 77(1):95–104. doi:10.1016/j.compag.2011.04.001

Zhang Y, Slaughter DC, Staab ES (2012) Robusthyperspectral vision-based classification for multi-season weed mapping. J Photogramm Remote Sens 69:65–73. doi:10.1016/j.isprsjprs.2012.02.006

Zwiggelaar R (1998) A review of spectral properties of plants and their potential use for crop/weed discrimination in row-crops. Crop Prot 17(3):189–206

Section III
Primary Weed Control Tools for Automation

Chapter 6
Precision Planting and Crop Thinning

Scott A. Shearer and Santosh K. Pitla

Abstract Historically, the management of inputs to crop production, especially seed, on agricultural lands has been controlled by humans using "field-average" practices. This chapter presents an overview of technology, available today and in the near term, that is altering the accuracy and precision of seed singulation and placement during planting. Increasingly, the cost of genetically modified organisms (GMOs) and biological and chemical seed treatments demands high-level accuracy and precision in seeding operations. Alternately, when it is impractical to singulate individual seeds because of seed size or shape, producers may choose to overseed a crop and then thin the resulting stand to achieve an optimal stand. In either event several enabling technologies are available to enhance the likelihood of success. Characteristics of the precision planting systems (tractor and planter combination) of the future will change in three key areas, including (1) precision seed meters and methodologies that deliver and position seed within the furrow, (2) individual row stepper motor drives that index seed placement in accordance with distance traveled, and (3) seed metering capabilities that support seeding of multiple varieties/hybrids within one field.

S.A. Shearer (✉)
Department of Food, Agricultural and Biological Engineering,
The Ohio State University, 590 Woody Hayes Drive, Columbus, OH 43210, USA

Biosystems and Agricultural Engineering, University of Kentucky,
128 C.E. Barnhart Building, Lexington, KY 40546-0276, USA
e-mail: shearer.95@osu.edu

S.K. Pitla
Department of Food, Agricultural and Biological Engineering,
The Ohio State University, 590 Woody Hayes Drive, Columbus, OH 43210, USA
e-mail: pitla.1@osu.edu

S.L. Young and F.J. Pierce (eds.), *Automation: The Future of Weed Control in Cropping Systems*, DOI 10.1007/978-94-007-7512-1_6,

1 Introduction

Historically, the management of inputs to crop production, especially seed, on agricultural lands has been controlled by humans using "field-average" practices. While most producers recognize that variability exists in seed supplies and seeding techniques, the tools to address this variability were lacking. The development of spatial management technologies such as the Global Navigation Satellite System (GNSS), geographic information system (GIS), and controller area networks (CAN) now permits agricultural resource management decisions to be made with greater specificity, precision, and accuracy. Site-specific management has been around for many years, but it was the tools developed in the 1980s and 1990s that have allowed this management method to grow in popularity and be extended to other areas of agricultural production. An important outcome is the opportunity to manage variability within a production unit of land at increasingly finer resolution when compared with the "field-average" approaches of the past. However, site-specific management is currently limited by the ability of existing equipment to physically implement specific input management strategies. For example, pneumatic seed meters are susceptible to a number of distribution and control errors which compromise their ability to effectively establish uniform plant stands in accordance with local soil conditions and seed lot variability.

2 The Evolution of Technology

At the turn of the previous century, the introduction of the farm tractor, the replacement for animal power, was met with skepticism. The principal argument of the day was whether or not agricultural producers could afford to own this new technology. Most producers realized that a significant amount of time and land resources were dedicated to caring for animal power sources. The internal combustion engine, an alternative power source, freed producers to concentrate on crop production and to be timely in the completion of field activities. The downside was that most farming operations became more capital intensive. However, the mechanization of agriculture freed the rural labor force to move to manufacturing-based careers in the cities thereby fueling the industrial revolution.

Today, with the advent of the semiconductor and associated development of embedded controls and sensing technologies, we see a continuing focus by agricultural equipment manufacturers on removing the human operator from the control loop. With the continuing development of the Global Navigation Satellite System (GNSS), a ubiquitous and affordable radio-navigation facility, we can now track and control field machinery operation to within centimeter-level accuracy and precision for horizontal positioning nearly anywhere on the surface of the earth. It is the marriage of our ability to precisely control the timing, rate, and placement of production inputs with the enhanced genetic potential of GMO crops that allows us to achieve production increases never thought possible. Increasingly, management of seed at planting is an important key to unlocking the genetic potential of the crop.

While the title of this chapter is *Precision Planting and Crop Thinning*, what most producers are after is stand establishment – knowing that they will have a spatial distribution of viable plants that does not compromise the yield potential of the crop. Stand establishment can be viewed from two perspectives: (1) precision planting of viable seeds or (2) overseeding with precision thinning (plant removal) to achieve the desired stand. With the first case the producer must have confidence in the quality of the seed and their ability to singulate, control spacing, and establish good seed-soil contact at an appropriate planting depth. Alternately, the seed can be metered at rates in excess of the desired stand and then thinned after the seed has germinated and the viability of the resulting plants has been established. The trade-off is cost – lost yield potential when the desired stand is not achieved versus additional seed and field operations. In either case seed placement and ensuring good seed-soil contact are essential elements of precision planting.

3 Seed Biology and Physical Properties

Critical to the success of any seeding mechanism is a fundamental understanding of seed biology and the physical factors important to successful germination. To begin one must develop a definition of germination which serves the purpose of the equipment designer. However, the definition of germination varies depending on the perspective of the agricultural professional. For example, the seed physiologist identifies seed germination as the "emergence of the radicle through the seed coat." Unfortunately, this definition falls short as it does not address the number of viable seedlings. On the other end of the spectrum, germination can be defined as the "emergence and development of a seed embryo… indicative of the capacity to produce a normal plant under favorable growing conditions." For equipment designers and end users, the latter seems to be a more practical definition in that it helps to better define success when it comes to seeding practices.

From McDonald (2008) we learn that two types of germination morphology are possible. In the first case (epigeal) the cotyledons emerge from the ground attached to the hypocotyls (e.g., beans). During emergence the rapidly elongating hypocotyls are arched providing the necessary forces to break through the soil crust. Alternately, for hypogeal germination, the cotyledons remain in the soil. For the latter case the epicotyl is the rapidly elongating structure that breaks through the soil crust (e.g., maize). In either case the cotyledons, or comparable storage structures, provide the energy to sustain plant emergence from the ground.

Once placed in the ground, seed requires an appropriate physical environment to promote germination (i.e., appropriate moisture, temperature, and gas levels). Moisture is essential to initiate metabolic activity in support of germination. Further, there is a critical moisture content level required for germination (e.g., 30 % for corn and 50 % for soybeans). Similarly, O_2 and CO_2 gas levels in the soil impact germination. Threshold levels of O_2 are required to initiate germination while elevated levels of CO_2 tend to retard germinations. As one might suspect soil bulk density becomes an important factor when considering the balance of moisture and gas required in support of germination. Temperature too plays a significant role in the promotion of germination. Typically, the optimal temperature is dependent on species and cultivars or hybrids. In general, optimal germination temperatures, or temperatures promoting the highest percentage of germination in a short period of time, range between 15 and 30 °C. Lower temperatures tend to slow the process while high temperatures can cause denaturation of proteins required for germination. In general higher-quality seed germinates over a wider range of temperatures.

Now that some ground rules have been established for favorable germination conditions, it is appropriate to explore a few additional considerations when looking at the seed-soil environment. Water uptake by the seed is typically referred to as imbibition. Good seed to soil contact as well as available soil moisture are essential for good stand establishment. Many factors affect the rate at which this process proceeds. These factors include seed coat permeability, seed composition, and soil moisture level. Above all, the most important controllable factor in seeding relative to moisture imbibition is seed-soil contact. Seed treatments (surface roughness) and seed size affect imbibition and can be altered to some extent. Alternately, furrow opening and closing, seed delivery, and soil tilth can be controlled, to some extent, by proper selection of cultural practices at planting. For example, extensive tillage prior to seeding may produce more uniform seed emergence when contrasted with no-till or conservation tillage practices.

4 Precision Seed Meters

Extended discussions of the development of precision seed meters are provided by Srivastava et al. (2006) and Heege and Billot (1999). These resources provide significant background and details of traditional seeding equipment for broad acre crops, including a typical pneumatic meter marketed in North America for row crops (Fig. 6.1). Unfortunately, the same resources are not available for vegetable and specialty crop seeding equipment. One of the better resources on modern seeding equipment for these crops is provided by Sanders (1997). In this extension outreach publication, the author overviews several seed metering devices. Three noteworthy devices, and a short description of each, follow:

Grooved Cylinder Style (*Gramore*) – This device requires round seed or coated seed that is made round. Seeds fall from a supply tube into a slot in a metal cylinder. The cylinder turns slowly and the seed drops out the diagonal slot at the bottom of the case. This meter has significant limitations and in general is not used with seed larger than peppers.

Belt-Type Style (*Stanhay*) – Circular holes are punched in a continuous belt at specified intervals. Seed is delivered by passing the belt through the seed mass to fill the cells (holes in the belt). Quality of singulation is best with spherical seed or seed made spherical through the addition of a seed coating. This technology is most appropriate for seed sizes ranging from tomato to watermelon.

Vacuum Style (*Gaspardo, Heath, Monosem, Stanhay, etc.*) – Air moves through holes in the periphery of a rotating disk singulating and trapping individual seeds against the metering plate. Excess seeds are removed with brushes and/or other mechanical means. Various models meter seeds ranging in size from lettuce to watermelon.

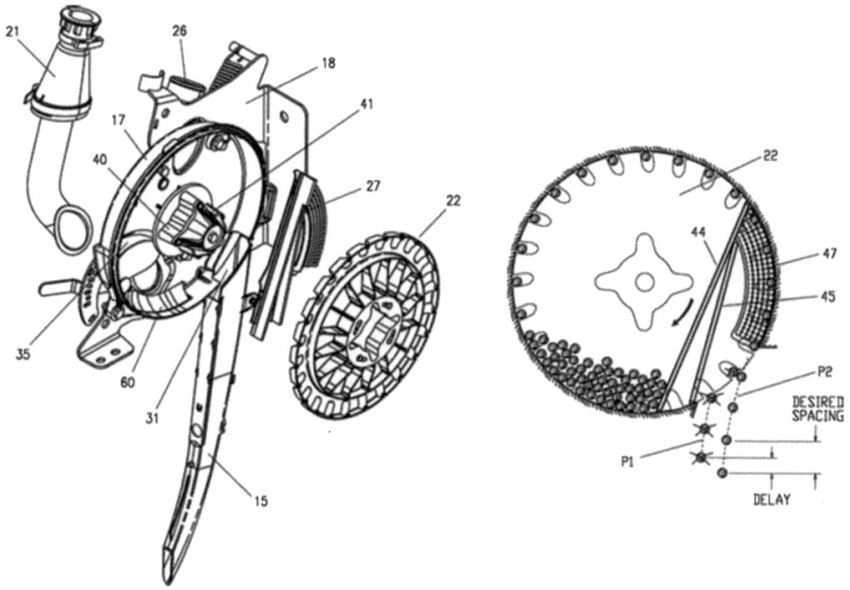

15 – Seed tube	35 – Sloping Back Wall
17 – Meter Housing	40 – Multi Lobe Disk Seat
18 – Mounting Bracket	41 – Multi-Lobed Clamp
21 – Air Release Assembly	44 – Primary Strip Brush
22 – Replaceable Seed Disk	45 – Secondary Strip Brush
26 – Air Inlet	47 – Seed Drop Brush
27 – Brush Assembly	60 – Sloping Surface
31 – Entrance of the Seed Tube	

Fig. 6.1 Seed meter sketch from US Patent No. 8375873 "Seed Metering Device for Agricultural Seeder"

5 Field and Zone Shape Management

Increasingly, agricultural producers around the world are turning to larger and faster equipment to be timelier in their operations. For example, in North America it is common to see seeding equipment that ranges up to and beyond 10 m in working width, sprayers that exceed 25 m, and grain harvesters exceeding 5 m. This focus on ever-increasing machinery size poses a serious problem as we consider the capability of this equipment when it comes to addressing the inherent variability of agricultural lands.

Many agricultural regions of North America, Europe, and developing nations include production sites with numerous small and irregularly shaped fields.

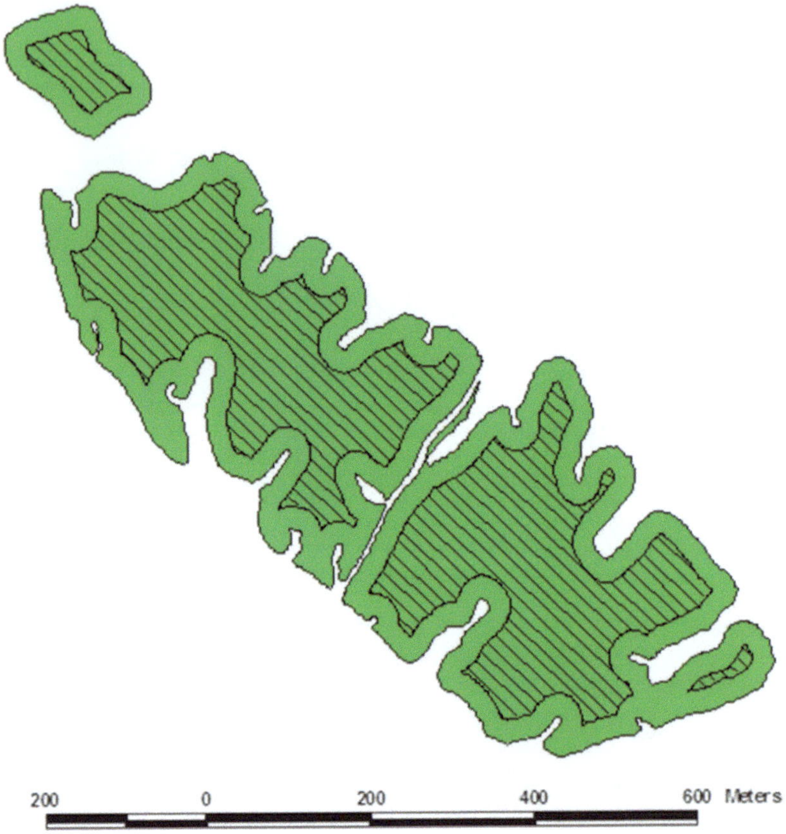

Fig. 6.2 Central Kentucky field with a total area of 40 ha. Two 24-row planter passes around the boundary of the field leaves 10.6 ha unplanted – approximately 50 % of the field is planted while turning

Typical planting practices when using large equipment require two passes around the periphery to establish the border rows. The interior region is planted with parallel passes. An example of a type of field found in Central Kentucky has an interior region of less than 50 % percent of the total (Fig. 6.2). Significant population variations occur across the planter width as the field margin is planted while turning. This is attributed to the outside rows dropping seed at the same rate as the inside rows while traveling at a significantly higher ground speed. The population across the planter width may vary as much as 100 % in tight turns, thereby wasting valuable resources such as seed, fertilizer, and pesticides. Population problems are further compounded when planting the point rows at the field margins. Equipment operators must determine if they will leave part of the field unplanted or plant into the border rows. In the latter case seed is wasted, and in some cases a yield reduction can be experienced in the double-planted regions. This problem is further compounded with subsequent spray and fertilizer applications.

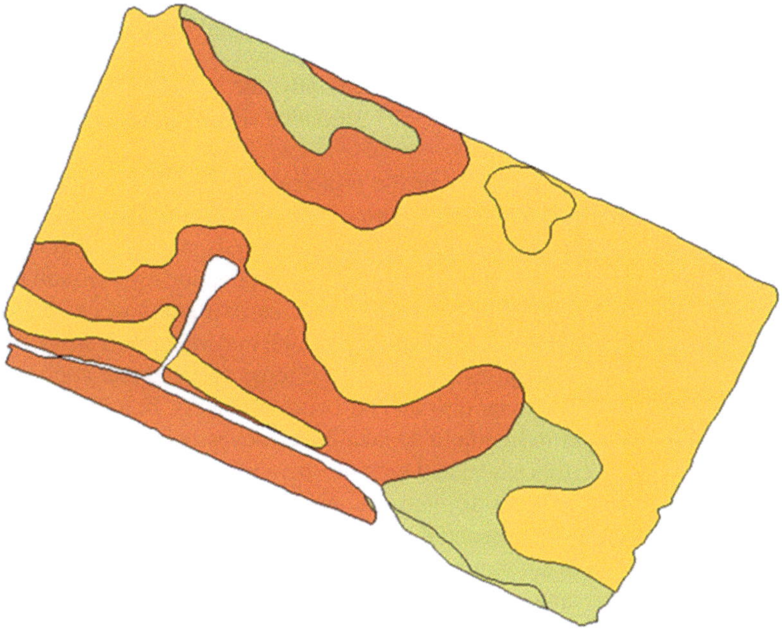

Fig. 6.3 Management zone delineation for a typical field in Central Kentucky using soil series as the basis. Six individual soil series were mapped to three seeding rates for producing maize grain

The delineation of management zones within agricultural field is becoming an area of study by itself (Fraisse et al. 1999). While the approaches vary significantly, there is one common thread – the shape and size of delineated regions vary significantly. What has prompted this track of research? In general, it is the realization that many factors contribute to variation. Some of the variation that exists can be traced to the underlying geology and processes controlling soil formation. Yet, other contributing factors include how this land was historically managed – pasture versus intensive row crop production.

Perhaps one of the more common approaches to describing this variability has been mapping soil series, which has occurred in the United States and many other countries around the world. While some might argue the utility of using soil maps for managing inputs for crop production, this example of a typical field soil (Fig. 6.3) sheds light on spatial management limitations that exist with equipment in the United States today. For clarity, the soil mapping units were clipped to the field boundary. And in this case the field boundary encloses only the cropped areas of the field excluding internal waterways and other grassed or forested areas not cropped.

While existing technologies continue to evolve, much of production agriculture in the United States is forced into "boom-width" management. Specifically, all variable-rate management is based on varying application or seeding rate across the implement width. With this limitation come several attendant problems that further degrade application accuracy. For most situations accuracy will be assessed as a summation of any deviation between the prescription map and actual application. With these definitions in mind, the following discussion explores several situations that contribute to "application error." Specifically the following issues will be addressed: (1) GNSS GPS accuracy, (2) guidance aides, and (3) variable-rate control of inputs.

Variable-rate fertilization was one of the principal driving forces in the development of precision agriculture. The basic approach was to first grid soil sample the field on a 1.0 ha grid, submit the soil sample to a lab for analysis, and then develop a prescription map based on a set of rules that tied nutrient application levels to soil nutrient levels. VRT application relies on the integration of several components to form an application system (GPS, task computer, controller and metering mechanism, and distribution). For illustration purposes we will first discuss variable-rate application of granular fertilizers using a spinner disk and air-boom spreaders. In follow-up we will use an agricultural sprayer to highlight application errors associated with application over- and underlap and errors associated with the increasing application width of new sprayers. At the onset of these discussions, we recognize that proper setup and operation of this equipment is essential to minimize application errors. However, as with any technology, there are limitations to overall system performance.

As we learn how to manage the variability that exists within agricultural production units, as equipment continues to grow in size and speed, and on-the-go sensing technologies evolve, a new class of agricultural equipment will be required. Returning to the mid-1990s in the United States, the farm press introduced precision agriculture as "farming by the square foot." However, this has never really been the case in mainstream agriculture. Perhaps a more accurate statement, even today, would be "farming by the boom-width."

The first commercial offering of "farming by the square foot" technology can be credited to Solie et al. (1996) with the introduction of their combined NTech sensor and application control system. Originally, this system was developed to control the application of N to wheat based on reflectance sensing of crop N stress. The basis of this system is a reflectance sensing element coupled with single nozzle metering and application of N. Once calibrated, this system is operated in real time. Data is shared between sensor and site verification data is logged using CAN. Benefits accruing to the users of this technology include a significant reduction in N application (up to 30 %) with little impact on final yield.

Perhaps a more universal field operation that will benefit through the application of distributed control is seeding. Seeding technology has changed significantly over the last 70 years. The old cell style meters have been replaced with pneumatic meters. However, drives for all seeding metering mechanisms have remained the same – ground driven via roller chain. As the width of machinery continues to increase, seeding rates are still controlled via a ground-driven common shaft constraining today's planter designs.

Much of today's agricultural field machinery is designed for large agricultural regions. Contrasting the plains of Illinois with Central and Western Kentucky, there are significant differences in planting practices. In Kentucky no-till production is preferred and seeding must be accomplished under extreme conditions – high residue environments. High residue environments pose two problems: penetration for placing the seed below the soil surface and the ability to see the marks left by the marker arms. The latter problem causes confusion on the part of operators in that row spacing between adjacent planter passes is inconsistent and portions of a field are left unplanted or are double planted.

5.1 GNSS and Automated Guidance

The continuing integration of technologies such as GNSS, GIS, and CAN provide new opportunities for agricultural producers to better control metering and placement of crop production inputs (seed, fertilizer, and chemicals) during field operations. A significant driving force with these technologies is reduction in cost to a point where most components are affordable to producers of nearly any size. For example, agricultural-grade GPS receivers purchased in 1995 cost nearly $5,000 (US) with differential correction signal subscriptions of up to $800 (US) per year. The horizontal accuracy of these receivers was on the order of 2.0–3.0 m. Today, US producers can purchase receivers with free Wide Area Augmentation System (WAAS) correction for under $100 (US). The horizontal accuracy of these receivers is reported to be less than 2.0 m.

The single technology that now affords farmers the opportunity to manage variability is GNSS. Although this navigation technology existed for some time, it was not until the deployment in space that 24-h per day coverage became available to civilian users. In the United States, agricultural producers rely on the Global Positioning System (GPS) as deployed by the Department of Defense. It is nearly impossible to initiate a discussion on GPS with US farmers without discussing system performance and cost. Nearly all US producers recognize the trade-off between cost and accuracy. Unfortunately, the subject of GPS accuracy is not universally understood by end users. Most manufacturers focus on the positive attributes of their systems while minimizing unflattering performance attributes that may be important to end users. To this end it is essential to understand accuracy and precision within the context of GPS coordinate fixes. More importantly, when GPS and/or GNSS are deployed for field operations, what can the end users expect?

There are essentially four classes of GPS receivers – low cost (2–3 m horizontal accuracy), agricultural grade (submeter), dual frequency (decimeter), and RTK (centimeter). All four classes require some form of differential correction, and as accuracy increases, so does cost. In the final analysis, the end user must consider carefully their requirement and the cost they can bear. Often, these users rely on industry-generated horizontal accuracy data for their purchase decisions.

5.2 Static Versus Dynamic Accuracy

While static accuracy is a good first approximation of how the GNSS receivers perform in actual applications, the deployment of GNSS in agricultural is unique. In most applications the receiver will be moving (e.g., yield monitoring, machine

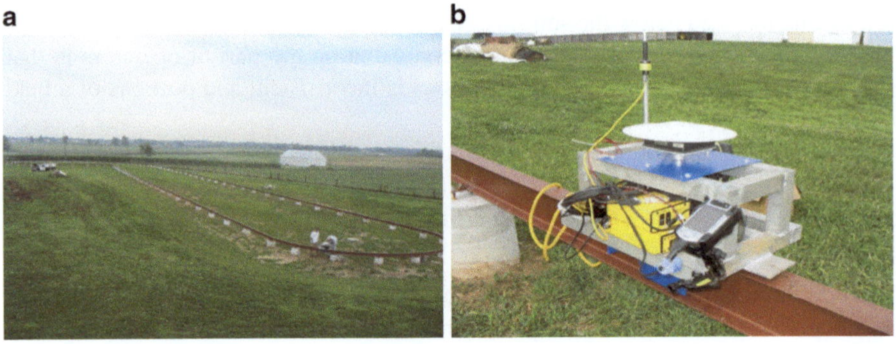

Fig. 6.4 (**a**) Continuous test track with 100 m straight parallel section and turns of varying radii and direction. (**b**) Test car with RTK rover and data logging device for extended testing (Stombaugh et al. 2005)

guidance, and variable-rate application). In reality the only static application of value to crop producers may be soil sampling. What, if any, are the differences between static and dynamic receiver accuracy for agricultural applications?

Perhaps the best approach to understanding dynamic accuracy in agricultural applications is to review test data collected by Stombaugh et al. (2005). The authors constructed a test track to address common situations that arise in agriculture (Fig. 6.4). The track consists of a closed, elevated I-beam with two 90 m parallel runs, a constant radius 180° turn, and four additional turns of varying radii (Fig. 6.5a).

Test data (Fig. 6.5b) were plotted from a receiver that was operated at a constant velocity in a clockwise (CW) direction. In turn the receiver appears to be averaging the position fixes, thereby creating an apparent receiver track that is skewed to the outside of the 180 degree turn. The error distribution seems to be an artifact of changes in receiver direction along the test track. In some application these errors may be of little consequence – such as for pass-to-pass guidance for spraying operations. For applications such as controlled traffic, where subsequent passes must be made in the same wheel tracks, and a variety of receivers are used, absolute errors may be unacceptable. If, in fact, these errors are systematic and repeatable, a simple offset correction may prove acceptable. Increasingly, the specification of dynamic performance is warranted given the demand for improved horizontal accuracy with today's equipment.

Horizontal "static accuracy" is the default metric reported with regard to GPS receiver performance and "accuracy" is "lack of error." The major problem encountered when evaluating GNSS receiver accuracy is knowing the "true" position for comparison purposes. For the accuracy values reported in manufacturer's literature, the end user must know if the accuracy being reported is "relative" or "absolute." "Accuracy" can be thought of in two parts: "precision" and "bias." "Precision" is the ability of the GPS receiver to produce the same position fixes (latitude and longitude) repeatedly when the receiver is in a fixed location. The difference between the position fix reported by the receiver and the "true" location is termed the "bias."

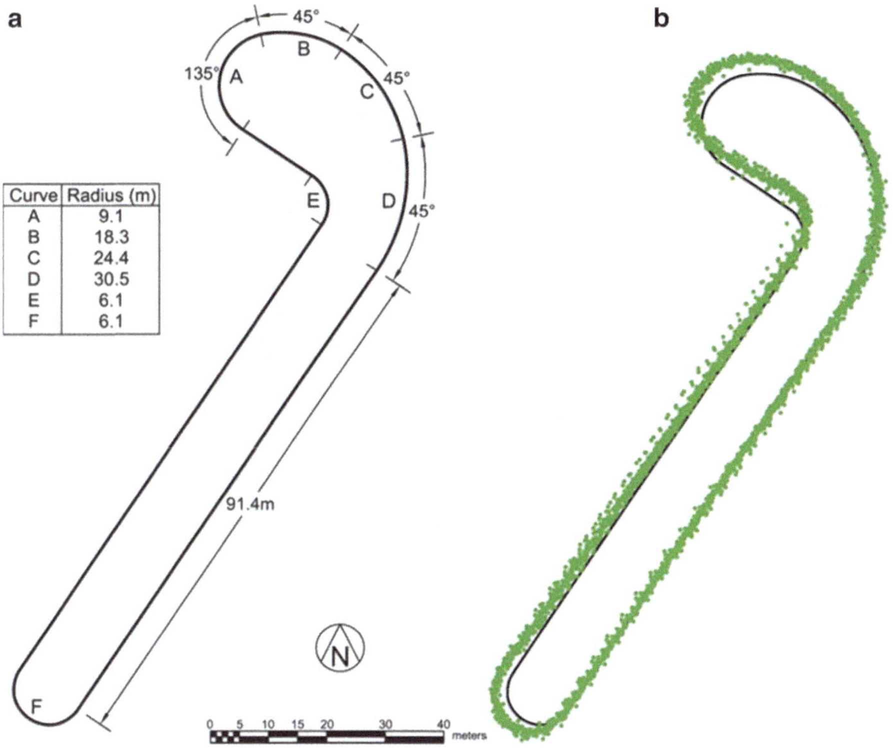

Curve	Radius (m)
A	9.1
B	18.3
C	24.4
D	30.5
E	6.1
F	6.1

Fig. 6.5 (**a**) Test track configuration with turns of five varying radii and 100 m parallel straight section for pass to pass receiver evaluation (Stombaugh et al. 2005). (**b**) Example test data set collected from a low-cost GPS receiver operated CW at constant velocity on the test track (Stombaugh et al. 2005) (Note: Directional dependence of deviations from test track centerline particularly in turns)

While the receiver may be very "precise," overall accuracy can suffer when a large "bias" exists. Are manufacturers reporting receiver "precision" with or without "bias"? Accuracy without the bias is referred to as "relative accuracy" while accuracy with the bias is termed "absolute." When managing inputs in accordance with zones, absolute positioning is essential for establishing the delineation between zones. For repeated field operations where production managers desire to control wheel traffic, control applications within zones, or plant into strip-till zones, knowing absolute receiver accuracy is essential.

The application of GNSS to guidance was first realized through the development of lightbars to assist the equipment operator in steering the vehicle to make adjacent, parallel passes at a predetermined distance. Acceptance of these devices was swift, in part, because of expanding equipment widths and difficulty experienced by equipment operators using foam marking systems (Wilkerson et al. 2003). The other selling feature of lightbars was the limitation of liability as operators were still required to steer the machine. More recently it was recognized that output from the

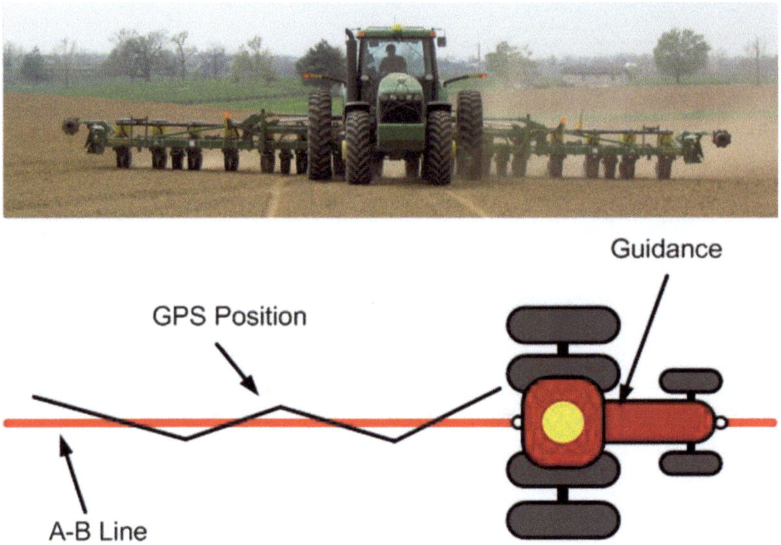

Fig. 6.6 Cross-track error determination for tractor guidance and implement following

lightbars in the form of guidance errors could be utilized in closed-loop feedback control to automatically steer the tractor. Essentially, this error signal is utilized to actuate a steering valve to guide the tractor along a predetermined path.

A large two-wheel drive (2WD) tractor with front wheel assist (FWA) was outfitted with a commercially available automatic guidance system and field tests were conducted (Veal et al. 2009). To study the motion of the tractor and implement, two RTK GPS receivers were fitted to the nose of the tractor and the center of the implement. The RTK GPS receivers used in this study achieved the same level of horizontal accuracy as the RTK system used with the automated guidance system. Position data were collected simultaneously for all three receivers and saved to a text file in a laptop. The tractor's path was set to allow for six concurrent 200-m-length parallel passes. Ground speed was set at 8.4 km/h for all test runs. Steering sensitivity settings were selected in accordance with implement type and draft load per recommendations published in the in the user's manual. The field implements used in this study included a 16-row towed planter, 16-row integral planter, and 4 m wide secondary tillage tool. Position data were collected from three RTK receivers and were converted from latitude-longitude (WGS84) to UTM coordinates then rotated to simplify the error calculations. A virtual A-B line was projected for each pass based on implement spacing (see Fig. 6.6). Cross-track errors (normal distance from traveled path to the projected A-B line) were determined for the logged data from each receiver. ANOVA (analysis of variance) tests of the error data were conducted to determine which factors influence the magnitude of cross-track error.

It was concluded that (1) the automated guidance systems steered the tractor along predetermined straight path (A-B line) with cross-track errors ranging up to

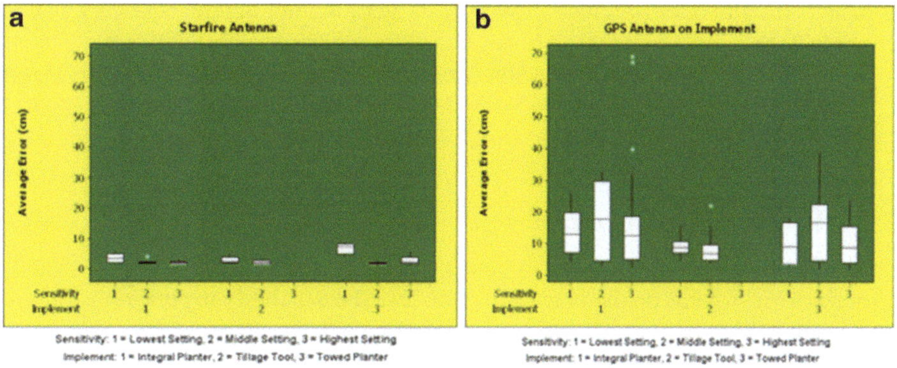

Fig. 6.7 Cross-track errors for tractor guidance (**a**) and drawn implement (**b**)

12 cm and a mean error of 3.21 cm; (2) tracking of the implement along the A-B line was not as accurate; (3) depending on soil conditions, slope, and steering sensitivity, the implement cross-track error may be 10 times greater than the cross-track error calculated at the receiver location on the tractor; (4) cross-track errors for the integral and towed planters were similar; and (5) steering sensitivity was found to influence cross-track errors.

The implement cross-track error clearly has the greatest variability and the highest average error for a given implement and steering sensitivity (Fig. 6.7). Also, it appears that the tillage tool and the towed planter appear to trail the tractor along the A-B line better than the integral planter. This data supports the concept that greater mechanical impedance created through the implement/soil engagement improved tracking accuracy and stability of the trailing implement. The data also supports that improved implement stability translates into improved overall system performance as it appears the tractor and tillage tool produced the most accurate A-B line tracking scenario. The box plot for the tractor receiver (top figure) has the least amount of variability and the lowest average cross-track error (typically less than 4 cm). Also, the variability of the cross-track error is smaller for steering sensitivities in the middle of the suggested range.

Many producers are looking for solutions that ensure implements follow with similar horizontal accuracies. A solution gaining in popularity is the addition of a second complete guidance package for the implement. In the case of soil-engaging implements, the steering mechanism is actually two or more large-diameter straight coulters mounted on kingpins and steered by a hydraulic cylinder. In turn the kingpins are mounted to the tillage implement via a bolt-on subframe. The coulters penetrate the soil surface generating sufficient lateral forces to steer the implement and reduce cross-track errors. While the cost of implement and tractor guidance is double that of tractor guidance, the solution allows soil-engaging tools to track with the tractor. When using RTK GNSS, subsequent field operations can be controlled with high absolute accuracy and precision. Implement guidance makes subsequent mechanical cultivation in close proximity to germinating and/or emerging plants possible – thereby adding significant value to organic cropping systems which rely on mechanical cultivation (Sorensen and Jorgensen 2005; Katupitiya and Eaton 2008; Young 2010).

5.3 *Variable-Rate Control*

Ground drive systems on the planters do not have the capability to perform variable-rate seeding. Hydraulic drive systems are used to vary the speed of the seed metering units to achieve variable-rate seeding. These drive systems consist of a hydraulic motor which is powered from selective (hydraulic) control valve on the tractor. Current day tractors have the capability to keep up with the flow rate demands of the hydraulic motor running the row unit seed meters. By varying the speed of hydraulic motor, seed population (seeds/acre) can be varied. A commercially available variable-rate drive system is used to drive the seed meters on a row crop planter, which allows the operator to program up to six seeding rates and then change seeding rates on-the-go during planting. The latter can be achieved through manual changes in seeding rate commands or automatically via map-based prescriptions. When a seed rate change is issued to the controller, the fluid flow rate to the hydraulic motor is altered. A speed sensor mounted on the motor sends a feedback signal to the controller confirming the change in the seeding rate. Additionally, motion sensors and potentiometers are installed on the planter frame to switch the row units ON and OFF when the planter is lowered and raised.

5.3.1 Section Control

Automatic section control enables wide implements to apply crop inputs (e.g., seed, fertilizer, and chemicals) across via user-selected section widths of the implement. Mechanical power directed to individual seed meters can be controlled either individually or by sections by installing mechanical clutches. The control sections are turned off when the planter passes through already-planted areas in the field avoiding double planting. The significance of section control is evident when the planter passes through point rows, waterways, and headland turns. Overplanting (doubles) and underplanting (skips) are avoided with section control, which translates to material input cost savings and improved yields, respectively. Planters with section control yielded an average savings of 4 % on seed costs and the savings increased for irregular-shaped fields with grass waterways and terraces (Fulton et al. 2011).

Section control requires a GNSS receiver, controller with enough channels to issue commands to the appropriate number of sections, and row clutches to engage and disengage the seed meters. North American equipment manufactures provide automatic section control systems with either pneumatic or electric clutches (Fig. 6.8).

Fig. 6.8 (a) Pneumatically actuated row clutch (Source: http://www.trimble.com/agriculture/tru-count.aspx). (b) Electrically actuated row clutch (Source: http://www.agleader.com/products/seed-command/sectional-control/)

5.3.2 Seed Drop Sensors

Seed drop sensors have become an integral performance monitoring aspect of nearly all agricultural planters. Initially developed to provide equipment operators with feedback regarding proper planter operation (e.g., planter boxes that are out of seed, plugged seed meters), these devices now provide important feedback on the accuracy and precision of seed drop and spacing. Increasingly, producers rely on the signal produced from this sensor to evaluate the overall performance of the single most important aspect of their farming operation. Seed monitoring sensors are located on the seed tube to count the seed that is dropped from the planter row units (Fig. 6.9). This sensor allows the operator and the planter controller to diagnose problems with the seed metering units by identifying skips and doubles. Typical seed tube sensors can be of optical type or radio wave based. Optical-type sensors sense the shape of the kernels that are dropped, whereas the radio wave-type ones sense the mass of the kernels using high-frequency radio waves. Given the dusty environment the planters work in, radio wave-based sensors perform better as they are not prone to dust. Since the wave-based sensors are measuring the mass and not the shape, they can accurately differentiate between single and double kernels.

5.3.3 Down Pressure Sensing and Control

Downforce on planter row units should be controlled to ensure ideal planting depth. Undulating field terrain offers varying resistance on the row units, and the planter has to adjust to these upward dynamic forces. Just enough force should be applied to keep gauge wheels on the ground and allow the double disk openers to drop the seed at the right depth. Excessive downforce can compact the soil, whereas a downforce less than the required can lead to shallow planting. Thus, proper downforce adjustments on planter row units become significant for improving yields, which includes a sensor that adjusts the depth for the gauge wheels (Fig. 6.10a).

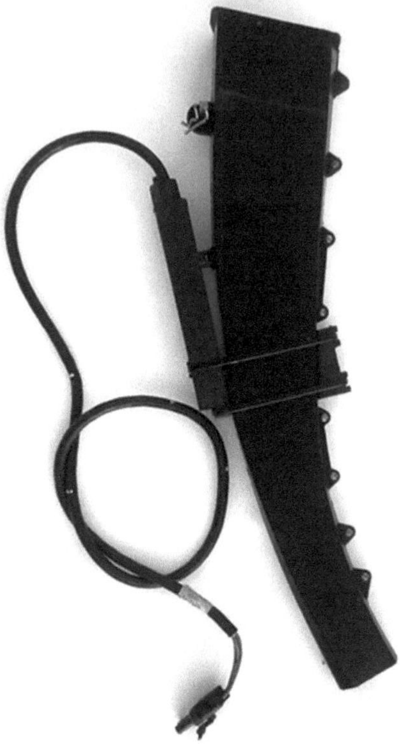

Fig. 6.9 Seed drop tube and sensor for assessing seed drop rate (Picture courtesy of S. Pitla)

Downforce can be applied either hydraulically or pneumatically and can be made automatically corresponding to the undulations in the ground during planting to place seed at the right depth. Hydraulic systems have a faster response time in mitigating the uneven terrains relative to the pneumatic systems. A hydraulic downforce system uses a hydraulic cylinder to put downforce on individual row units based on the feedback obtained from the gauge wheel sensor (Fig. 6.10b). The gauge wheel sensor measures the weight of the row units on the gauge wheels. In the pneumatic downforce system, the hydraulic cylinder is replaced by airbags that push the row units down.

5.3.4 Distributed and Embedded Controls

The deployment of electronics to agricultural field machinery was first met with skepticism in the mid-1970s. In North America the first agricultural electronics released to consumers were basic in nature often controlling only one or two machine functions such as bale tying and ejection for round balers. Since this time end users have come to appreciate the versatility this technology brings to the

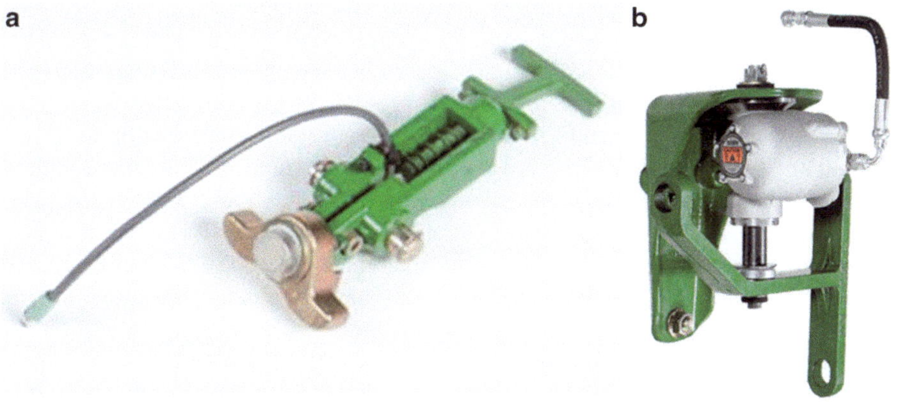

Fig. 6.10 (a) Planter down force sensor assembly (Source: http://salesmanual.deere.com/sales/salesmanual/en_NA/seeding/attachments/monitor_system/planters/seedstar_xp_row_unit_components.html) and (b) Hydraulic down-pressure actuator for agricultural planters (Source: http://www.dawnequipment.com/Dawn_Hydraulics.html)

control and adjustment of chemical application and seeding. In spite of the harsh environment (e.g., dust, vibration, corrosion chemicals, temperature extremes, moisture), controls are being added to modern equipment to achieve everything from emission reductions in off-highway diesel engines to complete adjustment of threshing and cleaning shoe settings when changing between harvesting of multiple crops and agricultural grain combines.

CAN-bus controls on agricultural tractors and implement are now commonplace. Demmel et al. (2001) reported on the use of the LBS DIN 9684 communications bus for accumulating field operations data. Erhl et al. (2002) investigated the effect of positioning system accuracy on data collected under the system proposed by Demmel et al. (2001). Darr et al. (2003) proposed a structure for tracking and reporting field operations. Central to this system was the use of CAN to accumulate data from an agricultural sprayer. The accumulated data were stored in a format that supported export to a variety of software packages for further analysis.

Implementation of CAN-bus communications is proceeding at a rapid pace in the United States. Stone et al. (1999) overviewed the progress of developing and implementing ISO 11783: An Electronic Communications Protocol for Agricultural Equipment. Implementation of control networks in agricultural machinery began in the mid-1990s. DIN 9684 and SAE J1939 provide the impetus and basis for developing ISO 11783. CAN 2.0B emerged as the favored message structure in ISO 11783 in part because of the 29-bit identifier when compared with the 11-bit identifier utilized in the LBS standard. The development and refinement of ISO 11783 continues to this day.

Implementation of the ISO 11783 has been reported by some researchers within the agricultural engineering profession. For example, Wei et al. (2001) employed CAN in the development of a distributed control sprayer. The sprayer was comprised of several reflectance sensing elements. Weeds were identified using color reflectance indices. Weed identities were passed to the sprayer control system via a CAN bus. Bus information was integrated with GPS data for the purpose of developing weed and spray application maps.

The current availability and low cost of microcontrollers coupled with the development of CAN communications are making distributed control of agricultural field machinery a reliable and cost-effective reality. CAN technology was first implemented in the automotive industry. Increasingly, with Tier II and III emission requirements, nearly every diesel engine manufactured today is equipped with CAN-based controls. Obviously, these actions have facilitated the move to off-highway use of the same technology.

CAN communication relies on a voltage differential between two wires – one that is high (CAN_H 3.5 VDC) and the other that is low (CAN_L 1.5 VDC). It is the time-varying voltage differential between these two wires that enables robust communications between devices with excellent immunity to signal noise. This two-wire bus is the physical layer of a CAN and has changed little since the initial development of CAN protocols. Data transfer rates of up to 250 Kbits/s are possible with CAN 2.0B serial buses (ISO 11783).

CAN is a message-based protocol unlike more traditional bus communications that are addressed based. Messages are transmitted to all devices (ECUs) on the bus. Embedded within each message is the priority of the message and the data being transmitted. Each ECU receives every message and then must decide if any action is warranted or if the message should be ignored. Alternately, each ECU may request information from any other ECU via a remote transmit request (RTR). The flexibility of this protocol permits ECUs to be added to the bus without reprogramming any of the existing ECUs thereby enhancing system expandability.

Under CAN 2.0B (ISO 11783) 134-bit messages can be sent at a rate of 1,900 messages per second. For example, a 10 Hz GPS receiver (generating 10 messages per second) uses less than 0.5 % of the total bus bandwidth. Further, message latency rarely exceeds 0.5 ms. The bottom line is that for agricultural applications where update and control rates of 1 Hz are common, the CAN 2.0B bus has significant bandwidth for most field activities even when the recommended limitation of 30 % of the total bus capacity is observed.

5.3.5 Electric Seed Meter Drives

Recent commercial offerings of similar, single-meter, electric drives were introduced in 2013 by US and EU manufactures. For the US manufactures, an internal ring gear was added to the pneumatic seed meter plate. This ring gear is driven via a 24 VDC electric motor with pinion gear (Fig. 6.11a). During the same cropping season, a German manufacturer (Fig. 6.11b) introduced a precision planter for the US market which utilized electric drives with a substantially smaller diameter seed meter when contrasted with other meters in the marketplace. The European planter required 90 A (12 VDC) of current to supply 24 seed meters. At the time of this

a b

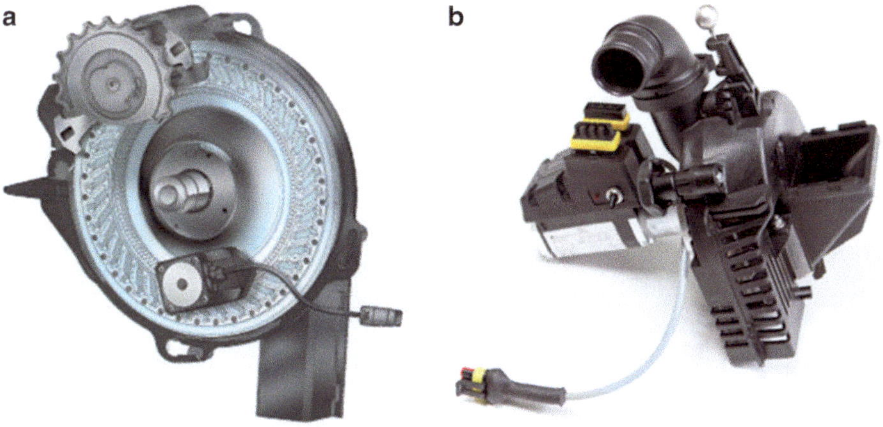

Fig. 6.11 Precision seed meters with electric drives: (**a**) Kinze Manufacturing (US) product offering (Source: http://www.kinze.com/feature.aspx?id=593&4000+Series+Vacuum+Meter) and (**b**) HORSCH Maschinen GmbH (Germany) product offering (Source: http://www.horsch2.com/en/products/seeding-technology/single-grain-seed-drills/maestro-cc/)

publication, the US-based manufacturer was able to control seeding rates on a row by row basis.

From unpublished work conducted by the authors, the following example highlights the development of robust, CAN-based, control system for seeding equipment (Fig. 6.12). This control scenario is the missing link needed for successful implementation of high precision seeding in North America. However, the justification for CAN-based control is multifaceted. The focus of this work was to (1) demonstrate the merits of CAN-based control, (2) highlight existing prototype development in US universities, and (3) fully develop CAN-bus communications for seeding equipment. In reality this same technology can be extended to field operations that involve chemical and fertilizer application. Similar benefits will accrue to those that adopt this compliment of technologies for all field activities that involve the metering and placement of crop production inputs.

For seeding equipment the goal was to replace the traditional mechanical drives with electric drives to facilitate the concept of individual row control. A review of existing seed metering devices led to the selection of a DC geared motor that would be able to produce sufficient speed for driving seed meter under actual field conditions (ground speed of 12+ km/h and populations of 90 K seeds/ha.) while providing adequate torque at low speeds. For prototype development modern, eight-row planter was selected. The motor utilized for this application was a permanent magnet 12 VDC motor with a worm gear right angle drive that produced 25 N.m of torque at speeds up to 70.0 rpm. The motor had a starting current of 7.0 amps and a continuous running current of 2.8 amps.

Speed control of the motor was accomplished using an H-bridge motor controller. The H-bridge allows bidirectional control of the motor as well as braking capabilities.

Fig. 6.12 Final prototype planter with GPS for generation of differential toolbar speed (**a**). Single row CAN node and motor controller (**b**). DC gear motor drive with sensor integration (**c**)

This particular controller was equipped with pulse width modulation (PWM). PWM is a digital square wave output with varying duty cycle or ratio of on-time to off-time. At high frequencies this signal becomes an average voltage output, percentage of full-scale voltage, which is very powerful in motor speed control. Motor speed feedback was needed to ensure precise metering of desired populations. This was achieved via an optical encoder. The optical encoder uses a combination of a light source, a rotary disk with evenly spaced windows, and a photodetector to measure angular displacement or angular speed. The initial prototype used optical encoder which generated 360 pulses per revolution of the output shaft of the motor drive. The processor selected to complete these tasks was an 8-bit microcontroller with 16-bit timer/counter, PWM, 32 K of FLASH memory, 1.6 K SRAM, 256 bytes of EEPROM, and a CAN engine. The microcontroller computes the desired motor speed from these variables and changes the PWM output until the desired rpm was achieved. The actual motor speed (10.0 Hz) and seed drop rate (1.0 Hz) were returned to the computer for logging.

Tests were completed to evaluate the ability of the system to precisely meter a desired population. System response to step input changes in ground speed was also

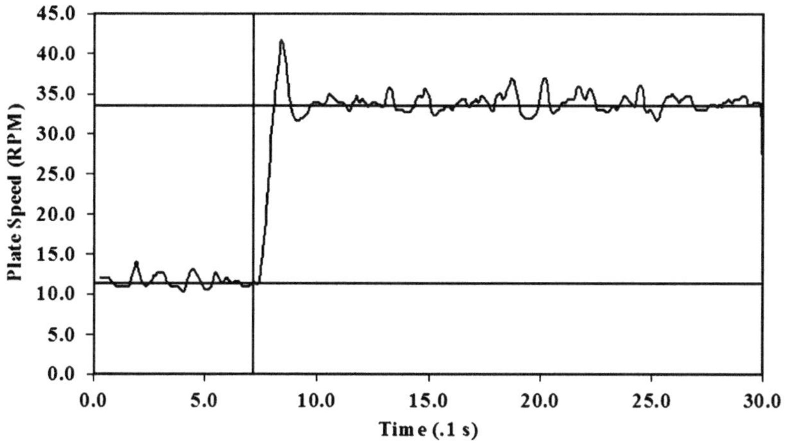

Fig. 6.13 Step response of initial prototype electric seed meter drive

evaluated. The 50 pulses per revolution provided by the encoder on the motor shaft produced 2,500 pulses per revolution of the output shaft. This combination allowed the system to sense 0.25 rpm difference at the output shaft at a 10.0 Hz sampling rate. Each seed meter was equipped with a control ECU, a motor controller, the motor/encoder combination, and a seed drop sensor. The system response to step inputs of the refined drive mechanism was quite accurate (Fig. 6.13). PID (proportional, integral, and derivative) feedback control was implemented to improve system response time, dampen speed oscillations, and improve steady-state errors.

Two, low-cost, GPS receivers were used to determine the speed differential across turning speed of the planter. The receivers utilized WAAS differential correction with a reported horizontal accuracy of 2–3 m. A receiver and a control ECU were placed at each of the outside rows and would read the velocity from the serial (RS-232) VTG NMEA string and send this over the CAN bus. Each meter ECU reads these two speeds to determine its velocity relative to its position on the planter. Map-based individual row control was accomplished using Windows-based task computer.

6 Crop Thinning

Thinning is the selective removal of seedlings or young plants to allow adequate space for the remaining plants to grow efficiently. In large-scale farming, techniques like precision seeding and transplanting can eliminate the need for thinning by starting plants at their optimum spacing. Unfortunately, for direct-seeded crops with small seed and poor germination, precision seeding may not produce desired results. For example, beets, carrots, onions, and other crops are often seeded at higher rates and then mechanically thinned to produce the desired plant density.

Fig. 6.14 Automated thinner prototype (Source: http://cemonterey.ucanr.edu/files/132403.pdf)

Crop thinning is the process of removing overpopulated plants to achieve desired yield and crop quality goals. Typically, the weakest seedlings in a row of crop are removed to create space for bigger and stronger seedlings. The premise is to reduce competition for sunlight, nutrients, and water intake to the strongest seedlings so that they can grow to their maximum potential. Crop thinning is done in forestry, direct-seeded row crops, specialty crops, and vegetable crops. Thinning operation is especially indispensable for crops like spinach, lettuce, cabbage, arugula, parsley, and cilantro where the leaf yield is crucial. Root crops like radish, carrots, parsnips, and sugar beet are also thinned to boost yields.

Manual thinning is physically challenging and monotonous and can cost up to $ 100 per acre in labor costs (Siemens et al. 2010). Crop thinning automation is seen as a way to mitigate labor shortages and high cost of production. Researchers are working on advanced crop thinning prototypes (Fig. 6.14), while some specialized thinning machines are already commercially available (Fig. 6.15). Automatic thinning is typically done either by mechanical cutting of the selected plant or by killing the plant using selective herbicidal spraying. In both cases, a machine vision system or an optical sensor is used to identify the plants to be removed from the row.

For example, an automatic thinner for lettuce consists of an onboard computer which houses the control algorithm to categorize the plants into the ones that need to be removed and the ones that need to be kept. The computer obtains signals from the machine vision system and provides ON/OFF control commands to spray the seedlings based on categorization. Raw images of the lettuce seedlings (Fig. 6.16a) are processed by the onboard computer (Fig. 6.16b). Based on seedling

Fig. 6.15 Crop thinning machine (Source: http://www.cemcoturbo.com)

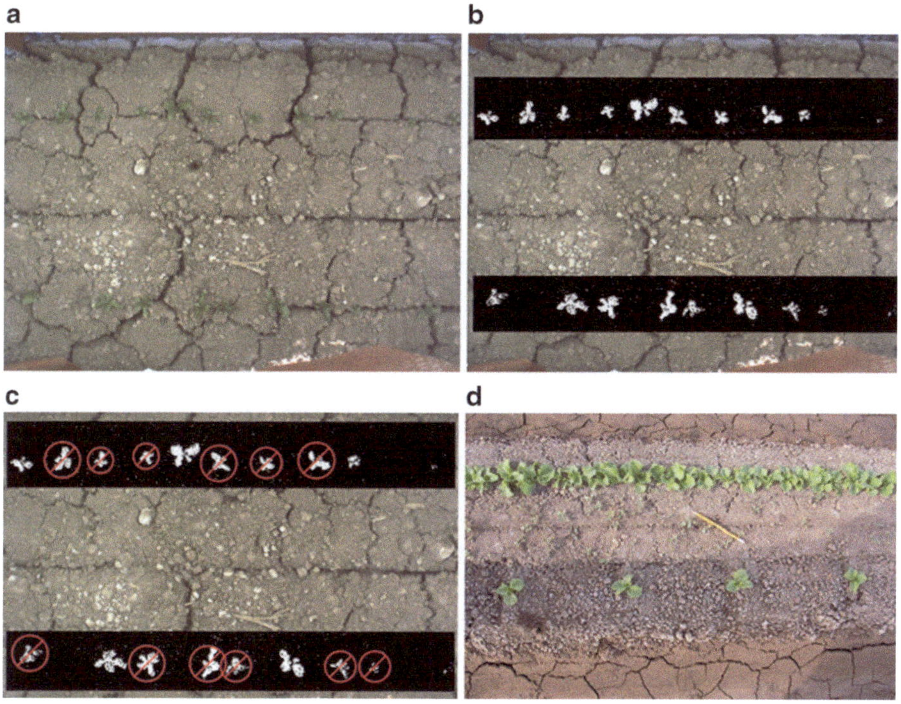

Fig. 6.16 Lettuce seedlings. (**a**) Raw image. (**b**) Processed image. (**c**) Plants to be eliminated are identified by the computer algorithm. (**d**) Thinned versus not thinned crop row (Source: http://cemonterey.ucanr.edu/files/132403.pdf)

Fig. 6.17 Row crop thinner (Source: http://www.agmechtronix.com/RCT.aspx)

characteristics, the computer algorithm identifies seedlings to be eliminated from the crop row (Fig. 6.16c). The automated thinning operation results in a thinned crop row (Fig. 6.16d).

Another commercially available row crop thinner works by selectively spraying chemicals for thinning (Fig. 6.17). This automated thinner can operate at ground speeds of up to 6.4 km/h. Selective spraying of the unwanted plants is achieved using computer vision and a touch screen controller mounted in the cab of the tractor.

7 Conclusions

The specification and addition of technology to stand establishment operations affords producers many options to achieve optimal crop performance. Increasingly, the cost of genetically modified organisms (GMOs) and biological and chemical seed treatments (Dyer et al. 2012; Munkvold et al. 2006) demands high-level accuracy and precision in seeding operations. Alternately, when it is impractical to singulate individual seeds because of seed size or shape, producers may choose to overseed a crop and then thin the resulting stand to achieve an optimal stand. In either event several enabling technologies are available to enhance the likelihood of success.

Pursuit of the optimal system for establishing desired crop stands begins with selection of the seed meter. As early as 1960 Mahoney (1959) recognized the need for "precision equipment for all phases of growing…" including precision placement of seed and fertilizer. When contrasting recent pneumatic seed meter development with the mechanical meters of the past, producers are now able to singulate a wider range of seed (hybrids, cultivars, and species) than ever before. The days of using fluted metering rolls for many crops (e.g., small grains, milo, canola, sugar beets) may be a thing of the past. However, seeding of some vegetable and specialty crops remains somewhat problematic with regard to small seed (e.g., celery, lettuce, radishes, brassica, onions). While meters for these crops continue to improve and

Table 6.1 Precision planting system of the future

Component	Characteristic
Tractor	Guided by automated guidance system and high-precision/high-accuracy RTK GNSS
Planter	Position and guidance system replicated from the tractor for implement guidance
	Pneumatics used to singulate and meter a steady stream of individual seeds
	Seed meters driven via individual stepper motors
	Embedded, CAN-based control for metering, singulation, and control of downforce levels on a row by row basis
	Seed drop and pneumatic down pressure sensors to monitor performance
	Residue management attachments for high residue and reduced tillage or no-till cropping systems
	Starter fertilizer attachments for placing nutrients in close proximity to seed and observing critical concentrations
	Openers and closing devices to prevent smearing furrow trench walls

seed producers do a better job of growing and processing quality seed, optimized yield may justify continued overseeding and thinning to desired plant stands.

Characteristics of the precision planting systems (tractor and planter combination) of the future will change in three key areas, including (1) precision seed meters and methodologies that deliver and position seed within the furrow while preserving seed orientation to optimize emergence uniformity and enhance plant canopy architectures (Torres et al. 2011); (2) individual row stepper motor drives that index seed placement in accordance with distance traveled to maintain desired plant densities and/or indexing of seed location between adjacent rows to desired crop canopy architectures; and (3) seed metering capabilities that support seeding of multiple varieties/hybrids within one field to match plant genetics with soil landscapes or multiple cultivars in polyculture systems (Fisher 2011). It is envisioned that these advanced planting systems will have robust and dynamic features that involve both the tractor and implement (Table 6.1).

References

Darr MJ, Stombaugh TS, Ward JK, Montross MD (2003) Development of a controller area network based handheld data acquisition system for identity preservation. ASAE paper no. 031103. Annual international meeting. Riviera Hotel and Convention Center, Las Vegas, NV, 27–30 July 2003

Demmel M, Rothmund M, Spangler A, Auernhammer H (2001) Algorithms for data analysis and first results of automatic data acquisition with GPS and LBS on tractor-implement combinations. In: Proceedings of 3rd European conference on precision farming in agriculture, Montpellier, France, 18–20 June 2001

Dyer A, Johnston J, Burrows M (2012) Small grain seed treatment guide. Montana State University, Bozeman

Erhl M, Demmel MR, Auernhammer H, Stepfhuber WV, Maurer W, Wunderlich T (2002) Spatio-temporal quality of precision farming applications. ASAE paper no. 023084. Annual international meeting. Hyatt Regency, Chicago, IL, 28–31 July 2002

Fisher M (2011) Do polycultures have a role in modern agriculture? *CSA News*, pp 4–10

Fraisse CW, Sudduth KA, Kitchen NR, Fridgen JJ (1999) Use of unsupervised clustering algorithms for delineating within-field management zones. ASAE paper no. 993043. ASAE international meeting, Toronto, ON, Canada, 18–21 July 1999

Fulton J, Brooke A, Winstead A, Ortiz B (2011) Precision agriculture series – timely information: Automatic Section Control (ASC) technology for planters (https://sites.aces.edu/group/crops/precisionag/Publications/Timely%20Information/Automatic%20Section%20Control%20%28ASC%29%20Technology%20for%20Planters.pdf)

Heege H, Billot J (1999) Part 1.3: Seeders and planters. In: Stout BA, Cheze B (eds) CIGR handbook of agricultural engineering, vol III: plant production engineering. ASAE, St. Joseph, pp 217–240

Katupitiya J, Eaton R (2008) Precision autonomous guidance of agricultural vehicles for future autonomous farming, ASABE paper no. 084687. ASAE, St. Joseph

Mahoney CH (1959) The canning industry's need of precision agricultural equipment. Trans Am Soc Agric Eng 1960:99–101, 104

McDonald MB (2008) Physiology of seed germination. The Ohio State University, Columbus

Munkvold G, Sweets L, Wintersteen W (2006) Iowa commercial pesticide applicator manual: seed treatment. Iowa State University Extension, Ames

Sanders DC (1997) Precision seeding for vegetable crops. Publication HIL-36, North Carolina Cooperative Extension Service, North Carolina State University, Raleigh

Siemens MS, Herbon R, Gayler RR, Brooks D, Nolte KD (2010) Automated machine for thinning, weeding and spot spraying (Source: http://cemonterey.ucanr.edu/files/132403.pdf)

Solie JB, Raun WR, Whitney RW, Stone ML, Ringer JD (1996) Optical sensor based field element size and sensing strategy for nitrogen application. Trans Am Soc Agric Eng 39(6):1983–1992

Sorensen CG, Jorgensen MH (2005) Intra-row weed control in organic crops – technical perspectives, capability and operational costs, ASABE paper no. 054146. ASAE, St. Joseph

Srivastava AK, Goering CE, Rohrbach RP, Buckmaster DR (2006) Chapter 9: Crop planting. In: Srivastava AK (ed) Engineering principles of agricultural machines, 2nd edn, ASABE. ASABE, St. Joseph, pp 231–265

Stombaugh TS, Cole J, Shearer SA, Koostra BK (2005) A test facility for evaluating GPS dynamic accuracy. In: Proceedings of the fifth European conference on precision agriculture Uppsala, Sweden, 8–10 June 2005, pp 605–612,

Stone ML, McKee KD, Formwalt CW, Benneweis RK (1999) An electronic communications protocol for agricultural equipment. ASAE paper no. 913C1798. Agricultural equipment technology conference, Louisville, Kentucky, 7–10 Feb 1999

Torres G, Vossenkemper J, Raun W, Taylor R (2011) Maize (*Zea Mays*) leaf angle and emergence as affected by seed orientation at planting. Explor Agric 47(4):579–592. doi:10.1017/S001447971100038X

Veal MW, Shearer SA, Stombaugh TS, Luck JD, Koostra BJ (2009) Automated tractor guidance and implement tracking error assessment. In: Proceedings of VDI-MEG conference of agricultural engineering (Nr.2060) 67th international conference Land.Technik – AgEng, Hannover, 6–7 Nov 2009. VDI verlag GmbH, Dusseldorf

Wei J, Zhang N, Oard D, Stoll Q, Lenhert D, Neilsen M, Mizuno M, Sing G (2001) Design of an embedded weed-control system using Controller Area Network (CAN). ASAE paper no. 013033. Annual international meeting. Sacramento Convention Center, Sacramento, CA, 30 July–1 Aug 2001

Wilkerson JB, Hart WE, Moody FH, Stombaugh TS, Morrow TF (2003) Evaluating operator feedback accuracy of row-guidance systems with GPS, ASAE paper no. 031012. ASAE, St. Joseph

Young SL (2010) Weed control in organic cropping systems, can automation fill the gap? Resour Eng Technol Sustain World 17(1):8–9

Chapter 7
Automated Mechanical Weeding

M. Taufik Ahmad, Lie Tang, and Brian L. Steward

Abstract One of the most important agriculture practices is to properly manage weeds because weeds negatively affect crop yield and quality. There are many types of commercial mechanical weeders that use the three main physical techniques: burying, cutting, and uprooting to controlling weeds. Two categories of mechanical weed control approaches, inter-row and intra-row mechanical weeding, are reviewed and discussed. Specifically, the most commonly used manual inter-row mechanical weeding tools are reviewed and compared according to their working principles. The more challenging area of intra-row mechanical weeding is reviewed, and manually operated intra-row mechanical weed control tools are compared. The current state of the art in automated mechanical weeding is discussed along with some cutting-edge technologies for intra-row mechanical weed control found in industry and the research community.

M.T. Ahmad
Mechanization and Automation Research Center, Malaysian Agricultural Research and Development Institute (MARDI) Headquarters, 43400 Serdang, Selangor, Malaysia
e-mail: Taufik@mardi.gov.my

L. Tang (✉)
Agricultural and Biosystems Engineering Department, Iowa State University, 203 Davidson Hall, Ames, IA 50011, USA
e-mail: lietang@iastate.edu

B.L. Steward
Agricultural and Biosystems Engineering Department, Iowa State University, 214D Davidson Hall, Ames, IA 50011, USA
e-mail: bsteward@iastate.edu

S.L. Young and F.J. Pierce (eds.), *Automation: The Future of Weed Control in Cropping Systems*, DOI 10.1007/978-94-007-7512-1_7,

1 Introduction

One of the most important agriculture practices is to properly manage weeds. Weeds affect crop yield due to competition to acquire plant nutrients and resources (Slaughter et al. 2008; Weide et al. 2008). Weeds have very fast growth rates compared to crops, and if not treated and managed, they may dominate the field. This competitive nature will unfortunately affect the crop yield (Slaughter et al. 2008). Gianessi and Sankula (2003) reported that most crops require that the field be kept weed-free during the first 4–6 weeks after planting to prevent serious yield losses from early season weed competition (Chap. 4).

There are many types of commercial mechanical weeders that use the three main physical techniques for controlling weeds: (1) burying, (2) cutting, and (3) uprooting. Burial of weeds is accomplished through the action of tillage tools (Gianessi and Sankula 2003) and is usually done during land preparation when soil conditions are enhanced through tillage. The goals of tillage include reducing the soil strength, covering plant residue, rearranging aggregates, and also removing weeds. The parameters to achieve a particular tillage effect are the ratio of forward speed to rotational speed, the diameter of tine rotation, the number of tines, the shape and design of tine tips, and the lateral offset to crop rows (Griepentrog et al. 2006). Cutting and uprooting weeds are performed by mechanical tearing and breaking the weeds from the soil and are usually done by mechanical cultivation after the crop is planted and has emerged. The majority of weed control implements are designed for use between crop rows (inter-row) (Cloutier et al. 2007). There are only a few machines that are designed for use in the intra-row of crops.

2 Mechanical Inter-row Weeding

Mechanical inter-row weeding is generally widespread and used by farmers in place of herbicides. The objective of inter-row cultivation is to cultivate as much of the inter-row area as possible without damaging the crop. Cultivation can destroy weeds by completely or partially burying weeds, uprooting, and breaking the contact between roots and the soil. However, there are limitations using this method. Weed control can only be done during the early crop stages because limited tractor and cultivator ground clearance and machine-plant contact may potentially damage the crop foliage at later growth stages (Cloutier et al. 2007). However, in spite of these limitations, there is a wide selection of cultivation implements that can be used for mechanical inter-row weeding.

Inter-row cultivators are the most common machine used for mechanical weed control. This agriculture implement consists of cultivating tools mounted on a toolbar that either rotate or sweep to move soil, bury, cut, or uproot the weeds. The sweeping-type cultivators use triangular-shaped or duckfoot-shaped blades that are swept under the soil but near the soil surface. The blades vary in width, from as small as 5.1 cm (2 in.) to as large as 71.1 cm (28 in.). This type of cultivator does not require

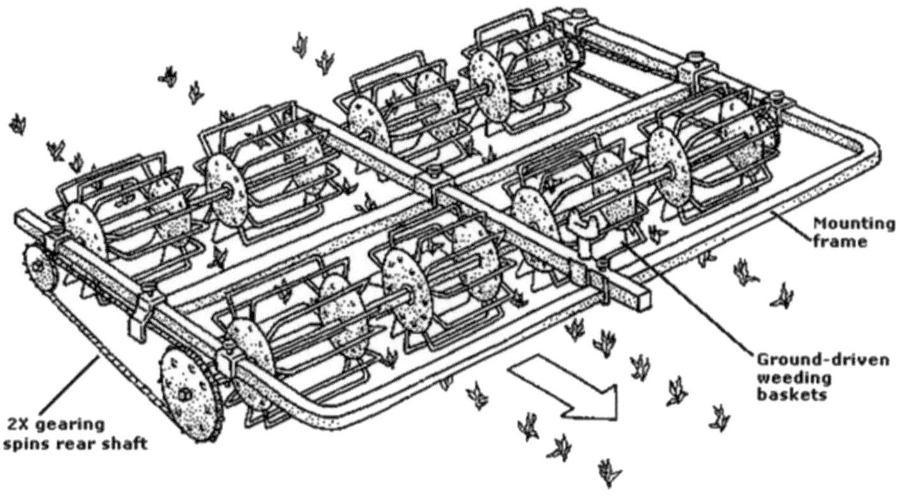

Fig. 7.1 Basket weeder for inter-row weed control (Bowman 1997)

any power takeoff (PTO) power, which means it does not require any power other than that provided through the draft force from the tractor. Recommended travel speeds for sweeping-type cultivators are 4–7 mile/h. Another type of cultivator is rotating type such as rotary tilling cultivator and rotary tiller, which are commonly used for inter-row weed control. However, the latter machine is more expensive, since it has been designed for multiple functions, such as strip planting into cover crops and preparing permanent plant beds. These rotary tilling implements use individually suspended inter-row gangs or blades, which are mounted on circular discs with parallel linkages. The cutting blades or knives vary in width, from 5–60 in, and in configuration. Metal housings can be used to cover the tilling blades to prevent crop damage. Recommended forward speeds for rotating-type cultivators are 2.5–5 mile/h (Bowman 1997).

The basket weeder is an implement that consists of rolling rectangular-shaped quarter-inch spring wire forming a round basket (Fig. 7.1). This basket weeder is ground driven, similar to the sweeping-type cultivators. The basket weeder will remove weeds at the top surface of the soil, without moving soil into the crop row. This machine is suitable in moist soils in minimal clay content. It performs weed control at forward speeds of 6.4 km/h (4 mile/h) to 12.9 km/h (8 mile/h) (Bowman 1997).

3 Mechanical Intra-row Weeding

Mechanical intra-row weeders control weeds within the crop rows. These types of implements accomplish their goal using two different approaches depending on the crop density. The first approach is to use selective machines or add-on tools that can perform weed control close to the crop, without damaging the crop itself. This

Fig. 7.2 Finger weeder uses rubber spikes that are pointed at an angle towards the crop (Weide et al. 2008)

approach does not require any sideways movement of the weeder. The second approach is to use machines that have weeding tools that move sideways to conduct weed control around the crop canopy. Below are some of the machines that have been reported to be effective in weed control.

3.1 Finger Weeder

The finger weeder is a simple mechanical intra-row weeder that uses two sets of steel cone wheels to which rubber spikes or "fingers" are affixed. The fingers point horizontally outward at a certain angle and operate from the side and beneath the crop row with ground-driven rotary motion (Fig. 7.2). The rubber fingers penetrate the soil just below the surface to remove small weeds. The finger mechanism performs best in loose soil and poorly in heavily crusted, compacted soils or where heavy residue is present. This type of weeder is effective against young weed seedlings up to 25.4 mm (1 in.) tall and interacts easily with well-rooted crops. The recommended operating depth is 12.7 mm (0.5 in.) to 19.1 mm (0.75 in.). The recommended forward speed to use with this weeder is 4.8 km/h to 9.7 km/h (3–6 mile/h). Alexandrou (2004) evaluated the finger weeder and obtained weed efficacy results of 61 % of the intra-row weeds killed in organic corn. A disadvantage of using this method, however, is that the tractor must be steered very accurately so that the finger mechanism can work as close as possible to the crop rows (Bowman 1997; Cloutier et al. 2007; Weide et al. 2008).

Fig. 7.3 Torsion weeder uses flexible coil spring tines to sweep the weeds (Weide et al. 2008)

3.2 *Torsion Weeder*

The torsion weeder is another machine available for intra-row weed control. Torsion weeders use a rigid frame that has spring tines connected and bent so that two short tine segments are parallel to the soil surface and meet near the crop plant row. This arrangement allows crop plants to pass through the tine pairs (Fig. 7.3). The coiled spring tines allow the tips to flex with soil contours and around established crops. These weeders have been shown to reduce weed densities to 60–80 %. Torsion weeders require very accurate steering with relatively low forward velocities and hence have a low working capacity. Accurate steering is required to avoid damaging the crop, since the tines operate very near to the crop. Torsion weeders are often used together with precision cultivators to perform efficacious weeding (Bowman 1997; Cloutier et al. 2007; Weide et al. 2008).

3.3 *Brush Weeders*

Brush weeders uses flexible brushes made of fiberglass or nylon rotated about vertical or horizontal axes. The brushes are rotated using hydraulic power from the tractor. These weeders mainly uproot but also bury and break weeds. A

Fig. 7.4 Vertical-rotating brush weeder uses hydraulics and requires an operator (Melander 1997)

protective shield or cover can be installed to keep the crop from being damaged. An operator is required to steer the brushes to cultivate as close and as many weeds as possible without damaging the crop plants (Melander 1997; Cloutier et al. 2007) (Fig. 7.4).

Fogelberg and Gustavsson (1999) investigated the use of a brush weeder for intra-row weed control in carrots and reported that the brush weeder was effective at early weed growth stages, specifically in the 2–4 true leaf stages. Forty-five to ninety percent of the weeds were uprooted using a working depth of 0.6 in. They concluded that the major mechanism of weed control obtained by brush weeding was uprooting, because brush weeding applies a greater uprooting force compared to the root anchorage force for the weed plants.

Kouwenhoven (1997) also reported on research investigating a brush weeder for intra-row weed control. In an experiment conducted in maize and sugar beet crops, it was determined that the best rotational speed for the brush weeders was 240–360 rpm with a forward travel speed of 1.2 mile/h. Results showed that brush weeding for maize was more effective than hand weeding. However, sugar beet plant damage was reported due to steering inaccuracy and fine soil created by the brushing effect. Combining this together with the moist weather conditions, it resulted in additional weed plant emergence after the weeding operation.

3.4 ECO-Weeder

The ECO-weeder is an intra-row mechanical weeder that is three-point hitch mounted and trails behind a tractor. It is driven by the power takeoff (PTO) of the tractor to drive a belt system that powers two discs with tines (Fig. 7.5).

Fig. 7.5 ECO weeder requires an operator to move rotating weeding mechanisms with tines (HCC 2011)

This machine is quite similar to the brush weeder described above, but uses a mechanical drive and does not require any hydraulic power. It is a good option for small production-scale vegetable growers because of its low price and low maintenance costs. The minimum tractor size needed to power the ECO-weeder is 14.7 kW (20 hp), and the PTO speed required is 540 rpm. It still requires an operator to move two rotating discs with vertically oriented tines in and out of the crop row. The forward speeds used by farmers are between 0.5–1.5 mile/h, and the rotation speed of the weeding element is estimated to be 150–300 rpm, similar to that of the brush weeder as reported by Kouwenhoven (1997). It was reported by the manufacturer that the ECO-weeder can save up to 60 % of weeding costs when compared to manual weeding due to the reduced labor requirements: two workers instead of eight workers (Univerco 2011). The weed control efficacy has not yet been reported.

4 Automated Technology in Mechanical Weeding

Automation technology has been applied to mechanical weed control that has resulted in a combination of manual and machine approaches. By using automation, a machine offers the possibility to determine and differentiate the crop plants from weed plants and, at the same time, remove the weed plants with a precisely controlled device (Bakker 2009). Slaughter et al. (2008) in a review on autonomous robotic weed control systems identified four core technologies needed for automated weed control: (a) guidance, (b) detection and identification, (c) precision in-row weed control, and (d) mapping. He also described several intra-row weed removal mechanisms for robotic actuation. One of the mechanical-based designs was using mechanical knives that can rapidly position in and out of the crop row.

Row guidance systems can use machine vision for crop row detection and/or global positioning systems (GPS). Machine vision has the ability to identify crop rows at travel speeds ranging from 1.6–6.2 mile/h and produces very small errors in identification, ranging from 4.7–10.6 mile/h. Meanwhile, GPS has the ability to provide a lateral positioning accuracy along the row with RMS error of 2.4 in and the maximum error distance of 13 cm (Slaughter et al. 2008). However, row guidance systems require that (1) the crop be planted using Real-Time Kinematics (RTK) GPS-guided planting system or (2) the crop rows be mapped using some type of geo-referenced mapping technique.

Detection and identification of weeds and the crop is very challenging to perform in real time. Weed identification techniques rely on machine vision systems and image processing techniques (Chap. 5) such as biological morphology, spectral characteristics, and visual structure. Steward and Tian (2005) used environmentally adaptive segmentation algorithm (EASA) to develop real-time machine vision weed detection for outdoor lighting conditions. Tang et al. (2000) used color image segmentation using a binary-coded genetic algorithm (GA) for outdoor field weed identification under different lighting conditions.

Precision intra-row weed control can use mechanical, chemical, thermal, or electrical approaches. Mechanically automated weed control such as the automated thinners uses mechanical knives that travel in and out of the crop row or use a rotating hoe that could be height adjusted (Astrand and Baerveldt 2002).

4.1 Examples of Automated Mechanical Weeders

Tillett et al. (2008) tested a weeding machine using computer vision to detect plants. This automated intra-row weeder used a rotating half circle disc that rotated to avoid contacting the crop plants during weeding. A camera was mounted centrally on the implement at a height of (5.6 ft.) looking ahead and down such that the bottom of the field of view was vertically below the camera, and the full width of the bed was visible over a length of approximately (8.2 ft.). The position of the plants along the crop row and their location relative to the rotating disc were detected using computer vision (Fig. 7.6). An experiment on a cabbage plot was conducted using an intra-row crop plant spacing of (1 ft.) and a forward velocity of 1.8 km/h (0.5 m/s). Weeding treatments were conducted at 16, 23, and 33 days after transplanting (DAP). The best results were obtained at 16 and 23 DAP, with 77 and 87 % reduction in the number of weed plants, respectively. However, after two weeks of subsequent weed regrowth and new germination, the number of weed plants after the 16 DAP weeding treatment was still reduced by 74 %, while a number of weed plants after the 23 DAP treatment were still reduced by 66 %. Under the experimental conditions, it was

Fig. 7.6 Automated weeder machine uses hydraulics to rotate semi-circle discs (Tillett et al. 2008)

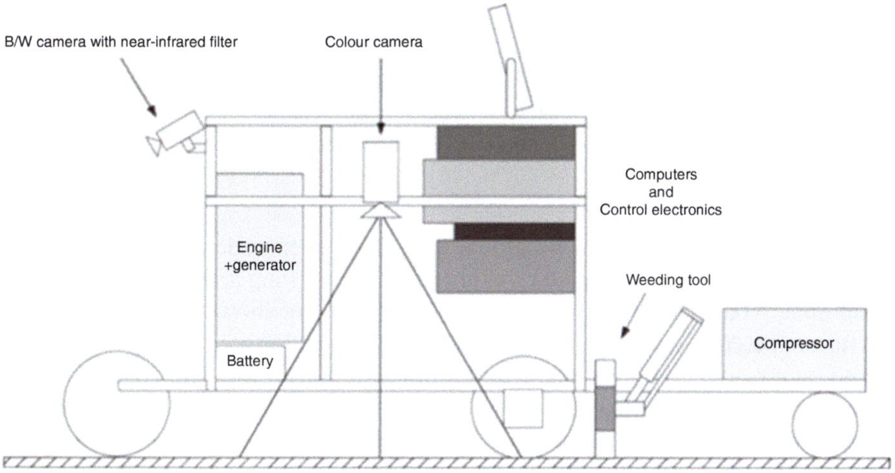

Fig. 7.7 Major components of the mobile robot (Astrand and Baerveldt 2002)

shown that performing weed control at an early stage succeeded in controlling later weed regrowth and new germination (see Chap. 4). This machine was commercialized under the name Robocrop (Inman 2010).

Astrand and Baerveldt (2002) developed an agricultural mobile robot with vision-based perception for weed detection and subsequent control. This machine required two cameras, one gray-scale camera with a near-infrared filter to obtain high-contrast images located at the front to identify the crop row location and direction and a color camera to identify crop plants, located at the center of the machine, facing downward toward the soil (Fig. 7.7). A weeding tool, which was a rotating wheel oriented perpendicular to the crop row, was located at the

Fig. 7.8 Sarl Radis intelligent weeder from France moves in and out of the crop row (Cloutier et al. 2007)

rear of the machine. The tool was lowered using a pneumatic cylinder when gap between crop plants was detected and provided some tilling action in the intercrop plant area. At a speed of (0.66 ft/s), the weeding robot showed good perception performance. The crop row detection camera was able to recognize crop rows based on a row-recognition algorithm with a (0.8 in.) error. The crop detection color camera successfully detected crops using image segmentation techniques to classify weeds and crops using color and shape features. However, the weed control efficacy of the machine was not reported. The research focused more on the perception system for crop row detection and crop detection and not on weed control in particular.

Cloutier et al. (2007) reported on the Inter-row hoe weeder developed by a France firm, Sarl Radis (Fig. 7.8). This automated weeder sensed reflected light from the field surface to detect crop plants and used a system to control the motion of a hoe around the crop plants. It was originally developed for transplanted crops and is best operated when the weeds are substantially smaller than the crop plants. This is usually the condition with conventional weeding, in which weeds are controlled while they are still small compared to the crop plants (see Chap. 4). The working speed of the prototype was reported to be (1.9 mile/h). FG (2008) reported that the Dutch Applied Plant Research organization is continuing to develop this prototype, hoping to achieve an operating speed of (2.5–3.7 miles/h) and to effectively control higher population weeds between the crops.

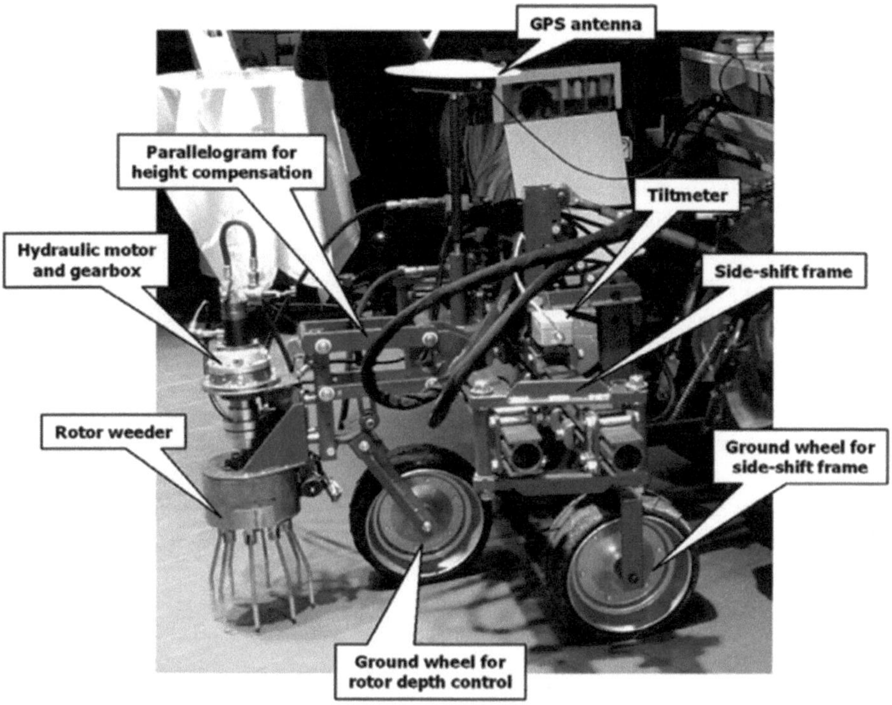

Fig. 7.9 Rotor tine weeder, also known as cycloid hoe, includes a side shift mechanism for lateral control and ground wheel for depth control (Griepentrog et al. 2006)

Griepentrog et al. (2006) developed an autonomous intra-row weeder based on RTK (Real-Time Kinematics) GPS to locate the weeder relative to crop seed maps that were developed at the time of crop seeding (Chap. 6). This weeder used a rotary weeding mechanism that is rotated using an electro-hydraulic motor. The mechanism consisted of eight tines with tine tips having an outer diameter of (0.77 ft.) (Fig. 7.9). These tines can be controlled individually to follow two different tine trajectories. The non-activated tine trajectories can be described as a cycloid curves, where a curve traced by a point on the circumference of a circle as the circle rolls on a straight line. The other trajectory is where the tine moves in and out of a crop row. The rotor weeding mechanism has the ability to control weeds inside the crop row and till the soil as close as possible to the crop plants without damaging them. The weeding effect of these tines is accomplished through uprooting, weed soil coverage, and root cutting. The machine was attached to an autonomous tractor driven using RTK GPS, and the lateral shift of the weed mechanism and the activation of the rotor tines were based on seed maps from the previous sowing operations.

5 Conclusion

Currently, most mechanical weeders range from 60 to 80 % reduction in number of plants, operate at depths ranging from 1 to 2 cm, and have a forward speed of 0.7–9.7 km/h. Automated weeding machines have not used electrical power for the weeding mechanism. Mechanical and fluid power has been widely used for controlling the weeding actuators. By using electric and electronics, it is hypothesized that more precise control of the weeding actuators can be accomplished. Also, the power consumption of the system can be monitored to understand the effect of soil depth, actuator speed, and other factors on required power. Electrical systems do not leak and cause soil contamination like hydraulics systems which is also prone to hydraulic fluid leakage.

Although the performance of current non-automated mechanical weeding technology seems promising, there are some other issues that should be considered. Machines such as the finger weeder and the torsion weeder require accurate steering to minimize crop damage. Brush weeders, although they have good performance, require an operator or operators at the rear to move the brushes in and out of the crop row. The more advanced vision-based weeders require slow forward speeds with a larger plant spacing to ensure good weed control. Automation is a natural next step for this concept since it has great potential to improve weed control efficacy.

References

Alexandrou A (2004) Evaluation of in-row weed cultivators in organic soybeans and corn. Organic Farming Research Foundation, Wooster

Astrand B, Baerveldt A (2002) An agricultural mobile robot with vision-based perception for mechanical weed control. Autonomous Robots 13(1):21–35

Bakker T (2009) An autonomous robot for weed control – design, navigation and control. PhD diss. Wageningen University, Wageningen

Bowman G (1997) Steel in the field: a farmer's guide to weed management tools, vol 2. Sustainable Agriculture Network, Beltsville, Maryland

Cloutier DC, Weide RY, van der Peruzzi A, Leblanc ML (2007) In: Upadhyaya MK, Blackshaw RE (eds) Mechanical weed management. CAB International, Wallingford, pp 111–134

FG (2008) SarlRadis Hoe. Farmers Guardian. Available at http://www.farmersguardian.com/sarl-radis-hoe/15587.article. Accessed 14 June 2011

Fogelberg F, Gustavsson AD (1999) Mechanical damage to annual weeds and carrots by in-row brush weeding. Weed Res 39(6):469–479

Gianessi LP, Sankula S (2003) The value of herbicides in U.S. crop production. National Center of Food and Agricultural Policy, Washington, DC

Griepentrog H, Norremark M, Nielsen J (2006) Autonomous intra-row rotor weeding based on GPS. CIGR World Congress, Bonn, Germany, p 7. Retrieved from http://www.staff.kvl.dk/~hwg/pdf/papers/Griepentrog2006b.pdf

HCC (2011) Eco weeder. Hillside cultivator company. Available at http://www.hillsidecultivator.com/?page_id=39. Accessed 8 July 2011

Inman JW (2010) Into gear: success of the robocropweeder/thinner. American Vegetable Grower. Available at www.growingproduce.com. Accessed 10 Apr 2010

Kouwenhoven JK (1997) Intra-row mechanical weed control–possibilities and problems. Soil Tillage Res 41:87–104

Melander B (1997) Optimization of the adjustment of a vertical axis rotary brush weeder for intra-row weed control in row crops. J Agric Eng Res 68(1):39–50

Slaughter D, Giles D, Downey D (2008) Autonomous robotic weed control systems: a review. Comput Electron Agric 61(1):63–78. doi:10.1016/j.compag.2007.05.008

Steward BL, Tian LF (1998) Real-time weed detection in outdoor field conditions. In: Meyer GE, DeShazer JA (eds) Proceeding of SPIE 3543, Precision agriculture and biological quality. SPIE, Bellingham, pp 266–278

Tang L, Tian LF, Steward BL (2000) Supervised color image segmentation by genetic algorithm for real-time weed sensing in outdoor lighting conditions. Trans ASAE 43(4):1019–1027

Tillett ND, Hague T, Grundy AC, Dedousis AP (2008) Mechanical within-row weed control for transplanted crops using computer vision. Biosyst Eng 99(2):171–178

Univerco (2011) ECO weeder. Available at http://www.univerco.net/cgi-bin/index.cgi?page=c2_2_0&langue=eng. Accessed 23 June 2011

Van Der Weide RY, Bleeker PO, Achten VTJM, Lotz LAP, Fogelberg F, Melander B (2008) Innovation in mechanical weed control in crop rows. Weed Res 48(3):215–224

Chapter 8
Targeted and Microdose Chemical Applications

Stephen L. Young and D. Ken Giles

Abstract In cropping systems, the precise application of herbicides is important for efficacious weed control. By using plant recognition and precision application technology targeting individual plants, off-target movement can be eliminated and herbicide rates significantly reduced without sacrificing yields. Highly targeted applications of nonselective herbicides into a growing sensitive crop are novel operations, nonexistent before the development of plant-specific targeting. New application technologies are essential when spatial rather than chemical selectively is to be deployed. In many potential applications, the chemical delivery system becomes the spatial resolution and speed limiting factor in the system.

1 Introduction

Weeds compete with crops for resources, including light, soil moisture, and nutrients. Significant yield reductions are associated with excessive weed growth and have been reported for all major crops (e.g., Donald and Khan 1992; Fischer and Ramirez 1993; Hall et al. 1992; Pike et al. 1990). Weed growth can be reduced with cultivation and cultural activities, including planting date, variety selection,

S.L. Young (✉)
Department of Agronomy and Horticulture, West Central Research and Extension Center, University of Nebraska-Lincoln,
402 West State Farm Road, North Platte, NE 69101, USA
e-mail: steve.young@unl.edu

D.K. Giles
Biological and Agricultural Engineering Department,
University of California, One Shields Ave., Davis, CA 95616, USA
e-mail: dkgiles@ucdavis.edu

S.L. Young and F.J. Pierce (eds.), *Automation: The Future of Weed Control in Cropping Systems*, DOI 10.1007/978-94-007-7512-1_8,

and cover crops in certain situations. The judicious use of herbicides is also an effective method for reducing weeds and, in the past half century, has been the primary tool in most crops grown on small to very large acreages. Lower efficacy is often associated with inadequate herbicide rates, development of resistance, improperly timed applications, or treatments that partially or completely miss the target. A repeat herbicide application is typically lower in efficacy, is expensive, and can have long-lasting effects on the weeds (e.g., weed resistance) and the environment (e.g., surface and ground water contamination).

Precision treatment of weeds utilizes ultralow doses of herbicides that are applied directly to the target at a very early life stage. By applying herbicides early in the life cycle of weeds, efficacy and crop yields can be improved significantly. Giles et al. (2004a) report 85–100 % control of pigweed species (*Amaranthus albus* L., *A. blitoides* S. Wats.), black nightshade (*Solanumnigrum* L.), and spotted spurge (*Chamaesyce maculata* (L.) Small) in newly planted tomato (*Solanum lycopersicum*) using a microdosing jet that delivered 37-uL per spray cell (0.63×1.25 cm). Similarly, Sogaard and Lund (2007) demonstrate a microdose system with a potential for controlling up to 100 weed seedlings m^{-2} using only 4 g ha^{-1}(12 ml ha^{-1}) of glyphosate. For 90 % control of yellow foxtail (*Setaria pumila* (Poir.) Roemer & J.A. Schultes) and velvetleaf plants, a direct application of glyphosate, using a mechanical end effector, required 22 % of the active ingredient (145 g aiL^{-1}) in a broadcast application (Hong and Tian 2009). Precisely placed herbicides can be very effective in controlling weeds without resulting in lower crop yields (Felton and McCloy 1992), but the commercial availability of precision application equipment is limited by its robustness in a wide variety of field conditions, including fluctuating weather and changing plant canopy and architecture (Moody et al. 2004). In addition, targeted recognition and application technology for precision weed control must be easily incorporated into current systems or used as stand-alone implements (Deng et al. 2010; Søgaard and Lund 2007).

Over the past decade, rapid advancements in automation and real-time recognition have occurred for weed control in cropping systems (see reviews by Singh et al. 2011; Slaughter et al. 2008a). The use of sensors and computers to quickly assess plants and their location within a field has led to the development of various systems. For example, a vision-based system was developed for broadleaf dock (*Rumex obtusifolius* L.) in grasslands using 2-D Fourier analysis to classify images (van Evert et al. 2009). Algorithms from the classified images successfully detected broadleaf dock in each image sequence covering an area of 1.5 m^2 every 30 milliseconds. In lettuce (*Lactuca sativa* L.), Slaughter et al. (2008b) used visible and near infrared reflectance spectroscopy to distinguish leaf and head lettuce varieties from weed foliage. Using equipment mounted on a mobile platform, 90 % crop vs. weed classification accuracy was obtained on over 7,000 individual spectra representing 150 plants. A machine vision-based detection system was used by Nieuwenhuizen et al. (2010) in sugar beet (*Beta vulgaris*) to identify and control volunteer potatoes (*Solanum tuberosum*) and had almost 80 % accuracy with very low crop death (1 %). The trend for improving plant recognition technology and incorporating it with other management applications (e.g., yield, soil nutrients, moisture) is increasing at a pace that is similar to the development of other high-end technology systems. For example, Zijlstra et al. (2011) describe technologically advanced devises, such as electronic noses that detect volatiles released by pathogens, acoustic detectors for identifying insects, and portable PCR units for real-time identification of fungal, bacterial, and viral diseases, as the future for monitoring pests in a comprehensive program for managing cropping systems.

While several research- and a few commercial-grade systems are being developed for targeted applications, little is known about the precise rates of herbicides that are needed to control very small weed seedlings. Similarly, little is known about the tolerance and recovery of crop plants when exposed to near proximal "micro-drift" rates of herbicide. Studies have been conducted on reduced doses and spray volumes (e.g., Schumacher and Hatterman-Valenti 2007), but not at the microscale. Dose-response relationships have been used most often for herbicide efficacy (e.g., Al-Khatib et al. 1995) and more recently for detecting herbicide-resistant weeds (e.g., Riar et al. 2011) and other less common weed control tools, such as flaming (Sivesind et al. 2009), clove oil (Boyd and Brennan 2006), and mustard seed meal (Boydston et al. 2011). More research is needed to evaluate the response of individual weed species to micro-rates and the efficacy of the equipment used for making targeted applications.

Similarly, little is known about the tolerance and recovery of crop plants when exposed to near proximal "micro-drift" rates of herbicide. Giles et al. (2004a) reported that "splash"-induced (i.e., "micro-drift") phytotoxicity experienced by the crop plants due to the micro-treatment reduced crop yield greater than weed competition from untreated weeds. The finding illustrated the potential for improperly executed micro-treatments to have a greater adverse effect than nontreatment of weeds. However, the work also identified the importance and usefulness of proper formulation, including physical property altering spray liquid adjuvants of the applied microdose treatments.

2 Efficacy of Chemical Weed Control

Weeds that have been injured by herbicides in the early growth stages (e.g., 2-leaf stage) are not likely to compete and survive in a field with a well-established and vigorously growing crop (Zimdahl 2004). Leaf sizes of weeds vary and can have a significant effect on herbicide coverage, which suggests that targeted applications can be tailored to meet specific individual plant sizes. What is true for broadcast applications of herbicides in identifying the precise plant growth stage that results in the most efficient and effective weed control treatment also applies to microdose herbicide applications made directly to plant surfaces.

The growth and development of weeds have been documented for many cropping systems (Buhler et al. 1998; Evans et al. 2003; Hall et al. 1992; Schier 2006; Wagner and Robinson 2006), showing the importance of implementing timely management strategies (see Chap. 4). A short period (e.g., 3–4 days) is sometimes all that is needed for plants to progress from cotyledon stage to the 2-leaf stage, demonstrating the ability of plants to quickly mature and thus the necessity for constant monitoring.

Glyphosate, a nonselective herbicide, is commonly applied to control annual weeds in cropping systems. In the field, a postemergence application at a typical field rate (1.6 L ha^{-1}) will kill many plants up to a certain growth stage. At this same rate, more mature plants are only injured and quickly recover. Early in the growth of a newly germinating plant, the surface of cotyledon leaves can vary from bare to very pubescent or hairy. As plants mature, the supple and malleable surface of seedling leaves increases in epicuticular wax content and becomes more resistant to absorbing liquids, such as herbicides and surfactants (Sanyal et al. 2006). Although the change in leaf surface texture is gradual in most weed species, it can play a role in limiting absorption and conductance of liquids across membranes (see Wang et al. 2007; Wang and Liu 2007). Therefore, the early stages of many weed species are the periods at which the leaf surface may be most likely to absorb an herbicide application, particularly at microdose concentrations.

In addition to texture, the role of leaf angle is a factor in limiting absorption and conductance of liquids across the leaf surface. The downward tilting of some weed species could be a response to the environment (e.g., sun, wind, rain) or an evolutionary response either to competition for light through a more aerodynamic and upright growth trajectory or a diversion of precipitation to the base of the plant (see Weinig 2000). For some weed species, the downward tilt of leaves makes it more difficult to get sufficient herbicide absorbed into the plant to cause death.

The traditional approach to making postemergence herbicide applications is through numerous nozzles spaced evenly along a boom that moves over the crop canopy. This method of applying herbicides emits an excessive amount of materials into the environment (e.g., off-target) where the target weed is located. The additional amount of materials can be easily quantified for comparison to microherbicide application rates. A 109 g $aeha^{-1}$ rate (1/8th of a typical field rate) of glyphosate in a microdose volume of 20 μl that is applied directly to the leaf surface of a velvetleaf weed in cotyledon-leaf stage requires 9.7 μg ae cm^{-2} to achieve over 90 % control (Young, unpublished data). The same rate of glyphosate applied in a typical spray volume of 187 Lha^{-1} would emit enough material to completely cover over 20 ha in a single layer of droplets ($187 \times 109 = 20{,}383$ g ae = 20,383,000,000 μg ae/9.7 μg ae $cm^{-2} = 2{,}101{,}340{,}206$ $cm^{-2} = 21.01$ ha). If the typical field rate were used (868 g $aeha^{-1}$), the area covered would quadruple twice to 168 ha (21.01 ha/0.125 or 1/8th of a typical field rate = 168 ha). Clearly, the excessive application of herbicides could be reduced with more targeted and precise applications.

3 Equipment for Chemical Weed Control

A fundamental performance demand for spatially selective, real-time treatment of weeds in close proximity with crop plants is the deposition of small volumes of spray liquid exclusively on the weed targets. This demand is a novel requirement, in contrast to conventional herbicide applications where the physical scale of target areas may be hectares. In spatially selective applications, within early season crops, the target areas may be on the scale of square millimeters, many orders of magnitude different from conventional, traditional herbicide application.

Achieving requisite, high spatial resolution of liquid deposit from a moving vehicle requires high-frequency, very brief emission times to reduce the minimum length of deposition along the axis of travel (i.e., the spatial resolution) and individual control of liquid emission sources to reduce the minimum width of deposition normal to the axis of travel (i.e., the spatial resolution). This demand for high-frequency delivery of small, repeatable volumetric doses to small spatial areas presents a unique design challenge. Typical agricultural nozzles are unsuited for this use due to their high flow rates and diverging spatial spray patterns. Because diverging fan spray patterns produce a spray width that is dependent on the distance from the nozzle to the target, variation in plant height or nozzle position above the plant would change the spatial resolution of the application. Additionally, the variation in droplet velocity across the sheet, when coupled with forward movement of the vehicle, alters the distribution of liquid deposit.

The demand for high temporal and spatial placement of a spray liquid (or any other weed control means) is a physical limitation. While sensing, detection, and navigation systems are continually being improved by advances in electronics and computer processing capabilities, physical placement and materials handling remain limited by physical actuators, positioners, and spray emitters. Physical systems are being improved, however, at a lower rate than electronic systems.

The current trend in design of the few microdosing systems that have been deployed in the field is to fabricate a liquid emitter source with a narrow treatment "footprint" that is the product of a physically narrowed width treated by an individual emitter and a rapid control means for actuating flow from the individual emitter. By creating a narrowly spaced array of individually controlled emitters and providing for high-speed on/off capability of each emitter, the "footprint" of each actuator, when in motion, is minimized, thereby creating high spatial resolution. In an example of this design, a common supply manifold is created to provide pressurized liquid to an array of high-speed solenoid valves that control the flow to individual "slices" of orifice plates (Fig. 8.1). The orifice plates create an array of cylindrical jets of spray liquid oriented vertically downward to the target area to be treated. The liquid jets provide a means to "spray" the target weeds while minimizing the deposition on nontarget crop plants. In this particular example, the manifold also provides the means for the liquid to be heated, allowing the potential for thermal treatment of weeds.

A typical pulsed-jet, microdosing system uses an array of circular jets oriented vertically downward. Circular jets have been theoretically and experimentally tested and found to be very efficient at retaining a high exit velocity for many diameters downstream. The width of the deposition area (dimension normal to the direction of travel) is determined by the number of individual jets joined into a distinctly controlled unit. The length of the deposition area (dimension along the direction of travel) is determined by the ground speed of the vehicle and the minimum pulse time of the jet array.

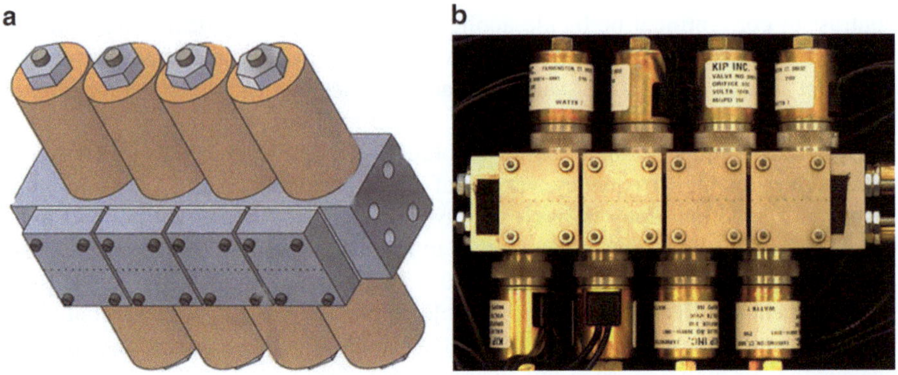

Fig. 8.1 Precision spray system for treatment of seedline weeds (Giles et al. 2004b). An engineering drawing rendered to show a manifold, control valves and orifice plates (**a**) A bottom view of the actual assembly showing the valves and the orifice blocks (**b**) (Photos courtesy of D.K. Giles)

The design criteria for a microdosing system is often established by the physical dimensions of the weed detection system with which it was intended to be used. For example, if displacement measurement along the direction of travel is a limiting factor, the resolution of the encoder for the ground wheel may be 0.65 cm (Lee et al. 1999). At a typical speed of 22 cm/s, which might be required for complex image analysis, the operating frequency (cell/s) will be 34 Hz with the frequency increasing proportionally with ground speed. In row crops, weed sensing systems can be limited to inspection and treatment of a narrow band (e.g., 10 cm along the crop row centerline). Outside of this band, weed control could be done by cultivation or continuous band spraying. Autoguidance systems, based on RTK GPS, may allow closer treatment, reducing the width of the band that must be inspected and spot treated. Lee et al. (1999) reported the development of a treatment system constructed as a linear array of type 304 W stainless steel hypodermic tubes, 1.25 cm long × 0.27 mm i.d. and inside chamfered on each end. Five tubes were placed 0.25 cm apart to create a linear array covering the 1.25 cm width. Eight, individually controlled arrays provided the 10 cm treatment width along the row centerline. Flow to each tube array was controlled by a direct-acting, DC solenoid valve with 12 V DC, 6 W coil, and 0.65 cm internal flow orifice. Minimum cycle time for the valve was measured as 6 ms; therefore, a minimum duty cycle of 20 % could be achieved at 34 Hz operation (Fig. 8.1).

An advantage of discriminating between plants to make spatially selective applications of herbicide to weeds is that nonselective herbicides can be used. This ability can reduce cost, improve chemical efficacy, and, when used in organic crop production, allow use of naturally derived, organic herbicides for weed control. However, inadvertent deposition of nonselective herbicide on the crop plant can result in significant phytotoxicity. This concern is important because the fundamental premise of a machine vision system is to allow weed control in the seed line, which is usually in close proximity to young crop plants that may be extremely sensitive to herbicide deposition. Often, the efficacy of foliar-applied herbicides is highly related to the uniformity of deposition and extent of leaf area covered by the spray deposit. Surfactants are commonly used with herbicide formulations to reduce surface tension and improve spread and uptake of the droplets after deposition. However, as surface tension is reduced, the potential for splatter and "splash" of the impacting high-energy jet increases. Even if the system was highly accurate at locating weeds and dispensing the liquid to exclusively strike the leaf surfaces of the weeds, the splash of the liquid stream could damage or kill the crop plants.

Giles et al. (2004a) and Downey et al. (2004) reported design and optimization of the fluid physical properties for use in highly resolved spatial treatment systems. The design demand was to engineer a fluid mixture that provided a high degree of target coverage and efficacy (consistent with a low surface tension and low viscosity) while preventing undesirable "splash" to target plants in close proximity to the target weeds (consistent with high viscosity and high surface tension). An optimal mix of surfactants and high molecular weight polymers was developed that provided an acceptable combination of efficacy and drift suppression.

Considering the future, as detection capabilities improve and the desire for higher vehicle speed increases, the demand for more highly resolved, both spatially and temporally, spray treatment systems will increase. Nonagricultural industries (e.g., high-speed printing, 3-D printing for fabrication) share the same design demand for improved spray actuators. It is likely that technologies developed for other demanding industries will be available for adaptation to this unique agricultural spraying need. Technologies such as piezoelectric actuators, ink-jet emissions, ultrasonic atomization, and robotic coating systems will offer potential solutions for agriculture.

An additional constraint, and perhaps a limiting factor beyond the technical concerns, is the regulatory status of allowing nonselective herbicides to be applied in fields with sensitive crop plants. Often a highly selective microdose application is made to a particular crop at a particular time in the season or in a particular location in a manner that is in conflict with the label instructions and chemical registration. Given that labels and regulatory registration data packages are developed and submitted by the chemical registrants with only traditional and conventional uses anticipated, the use of these chemicals in microdose, highly targeted applications may fall outside the intended use conditions. Therefore, the issue of legal status of the chemical use in these non-label applications will require resolution.

4 Conclusions

Precision application of weed control treatments requires sensitive technology that can track, record, and compute information on leaf shape, color, surface, and edge features for separating a weed and a crop plant (Hearn 2009; Meyer et al. 1998; Lati et al. 2011; Slaughter et al. 2008b; Tang et al. 2003; Tellaeche et al. 2011). The technology is still emerging and has a few challenges, including occluded leaves, misshapen leaves, moving leaves, and dusty leaves (see Chap. 15). Nevertheless, the algorithms to account for the changing plant and environmental conditions are being developed by engineers and computer scientists and will result in more accurate recognition and precision application systems (Zijlstra et al. 2011).

References

Al-Khatib K, Kadir S, Libbey C (1995) Broadleaf weed-control with clomazone in Pickling cucumber (*Cucumis-sativus*). Weed Tech 9:166–172

Boyd NS, Brennan EB (2006) Burning nettle, common purslane, and rye response to a clove oil herbicide. Weed Tech 20:646–650

Boydston RA, Morra MJ, Borek V, Clayton L, Vaughn SF (2011) Onion and weed response to mustard (*Sinapis alba*) seed meal. Weed Sci 59:546–552

Buhler DD, Netzer DA, Riemenschneider DE, Hartzler RG (1998) Weed management in short rotation poplar and herbaceous perennial crops grown for biofuel production. Biomass Bioenerg 14:385–394

Deng W, Zhao CJ, He XK, Chen LP, Zhang LD, Wu GW, Mueller J, Zhai CY (2010) Study on spectral detection of green plant target. Spectrosc Spectral Anal 30:2179–2183

Donald WW, Khan M (1992) Yield loss assessment for spring wheat (*Triticum aestivum*) infested with Canada thistle (*Cirsium arvense*). Weed Sci 40:590–598

Downey D, Giles DK, Slaughter DC (2004) Pulsed-jet microspray applications for high spatial resolutions of deposition on biological targets. At Sprays 14:94–109

Evans SP, Knezevic SZ, Lindquist JL, Shapiro CA (2003) Influence of nitrogen and duration of weed interference on corn growth and development. Weed Sci 51:546–556

Felton WL, McCloy KR (1992) Spot spraying. Agric Eng 73:9–12

Fischer AJ, Ramirez A (1993) Mixed weed infestations prediction of crop losses for economic weed management in rice. Int J Pest Manage 39:354–357

Giles DK, Downey D, Slaughter DC, Brevis–Acuna JC, Lanini WT (2004a) Herbicide microdosing for weed control in field-grown processing tomatoes. Appl Eng Agric 20:735–743

Giles DK, Slaughter DC, Downey D, Brevis-Acuna JC, Lanini WT (2004b) Application design for machine vision guided selective spraying of weeds in high value crops. Asp Appl Biol Pestic Appl 71:75–82

Hall MR, Swanton CJ, Anderson GW (1992) The critical period of weed control in grain corn (*Zea mays*). Weed Sci 40:441–447

Hearn DJ (2009) Shape analysis for the automated identification of plants from images of leaves. Taxon 58:961–981

Hong JY, Tian LF (2009) Direct application end effector for a precise weed control robot. Biosyst Eng 104:458–464

Lati RN, Filin S, Eizenberg H (2011) Robust methods for measurement of leaf cover area and biomass from image data. Weed Sci 59:276–284

Lee WS, Slaughter DC, Giles DK (1999) Robotic weed control system for tomatoes. Precisi Agric 1:95–119

Meyer GE, Mehta T, Kocher MF, Mortensen DA, Samal A (1998) Textural imaging and discriminate analysis for distinguishing weeds for spot spraying. Trans Am Soc Agric Eng 41:1189–1197

Moody FH, Wilkerson JB, Hart WE, Sewell ND (2004) A digital event recorder for mapping field operations. Appl Eng Agric 20:119–128

Nieuwenhuizen AT, Hofstee JW, van Henten EJ (2010) Performance evaluation of an automated detection and control system for volunteer potatoes in sugar beet fields. Biosyst Eng 107:46–53

Pike DR, Stoller EW, Wax LM (1990) Modeling soybean growth and canopy apportionment in weed-soybean (*Glycine max*) competition. Weed Sci 38:522–527

Riar DS, Norsworthy JK, Johnson DB, Scott RC, Bagavathiannan M (2011) Glyphosate resistance in a Johnsongrass (*Sorghum halepense*) biotype from Arkansas. Weed Sci 59:299–304
Sanyal D, Bhowmik PC, Reddy KN (2006) Leaf characteristics and surfactants affect primisulfuron droplet spread in three broadleaf weeds. Weed Sci 54:16–22
Schier A (2006) Field trials on weed control in genetically modified glufosinate resistant sugar beet under ploughing and reduced tillage conditions. J Plant Dis Protect 20:21–30
Schumacher CE, Hatterman-Valenti HM (2007) Effect of dose and spray volume on early-season broadleaved weed control in *Allium* using herbicides. Crop Prot 26:1178–1185
Singh K, Agrawal KN, Bora GC (2011) Advanced techniques for weed and crop identification for site specific weed management. Biosyst Eng 109:52–64
Sivesind EC, Leblanc ML, Cloutier DC, Seguin P, Stewart KA (2009) Weed response to flame weeding at different developmental stages. Weed Tech 23:438–443
Slaughter DC, Giles DK, Downey D (2008a) Autonomous robotic weed control systems: a review. Comput Electron Agric 61:63–78
Slaughter DC, Giles DK, Fennimore SA, Smith RF (2008b) Multispectral machine vision identification of lettuce and weed seedlings for automated weed control. Weed Tech 22:378–384
Søgaard HT, Lund I (2007) Application accuracy of a machine vision-controlled robotic microdosing system. Biosyst Eng 96:315–322
Tang L, Tian L, Steward BL (2003) Classification of broadleaf and grass weeds using gabor wavelets and an artificial neural network. Trans Am Soc Agric Eng 46:1247–1254
Tellaeche A, Pajares G, Burgos-Artizzu XP, Ribeiro A (2011) A computer vision approach for weeds identification through support vector machines. Appl Soft Comput 11:908–915
Van Evert FK, Polder G, van der Heijden GA, Kempenaar C, Lotz LA (2009) Real-time vision-based detection of *Rumex obtusifolius* in grassland. Weed Res 49:164–174
Wagner RG, Robinson AP (2006) Critical period of interspecific competition for four northern conifers: 10-year growth response and associated vegetation dynamics. Can J Forestry Res 36:2474–2485
Wang CJ, Liu ZQ (2007) Foliar uptake of pesticides – present status and future challenges. Pesticide Biochem Physiol 87:1–8
Wang SW, Duan LS, Eneji AE, Li ZH (2007) Variations in growth, photosynthesis and defense systems among four weed species under increased UV-B radiation. J Integr Plant Biol 49:621–627
Weinig C (2000) Plasticity versus canalization: population differences in the timing of shade-avoidance responses. Evolution 54:441–451
Zijlstra C, Lund I, Justesen AF, Nicolaisen M, Jensen PK, Bianciotto V, Posta K, Balestrini R, Przetakiewicz A, Czemborf E, Zandea J (2011) Combining novel monitoring tools and precision application technologies for integrated high-tech crop protection in the future. Pest Manag Sci 67:616–625
Zimdahl RL (ed) (2004) Weed-crop competition: a review, 2nd edn. Blackwell Publishing, Ames, pp 10–11

Section IV
Field Applications

Chapter 9
Field Applications of Automated Weed Control: Western Hemisphere

Steven A. Fennimore, Bradley D. Hanson, Lynn M. Sosnoskie, Jayesh B. Samtani, Avishek Datta, Stevan Z. Knezevic, and Mark C. Siemens

Abstract Opportunities for automated weed control vary widely among cropping systems of the Western Hemisphere. High-value conventional and organic horticultural crops may provide the best initial opportunity for automated weed control because of dependency on labor for hand weeding and the lack of effective herbicides. Fresh market vegetable crops are planted year-round in small successive plantings which make them an attractive target for weed control automation equipment.

S.A. Fennimore (✉)
Agriculture and Natural Resources, University of California, 1636 East Alisal Street, Salinas, CA 93905, USA
e-mail: safennimore@ucdavis.edu

B.D. Hanson
Department of Plant Sciences, University of California, One Shields Ave, Davis, CA 95616, USA
e-mail: bhanson@ucdavis.edu

L.M. Sosnoskie
Department of Plant Sciences, University of California, 1210 Plant and Environmental Sciences Bldg, Davis, CA 95616, USA
e-mail: lmsosnoskie@ucdavis.edu

J.B. Samtani
Department of Horticulture, Virginia Polytechnic Institute and State University, 1444 Diamond Springs Road, Virginia Beach, VA 23455, USA
e-mail: jsamtani@vt.edu

A. Datta • S.Z. Knezevic
Department of Agronomy and Horticulture, University of Nebraska, Northeast Research & Ext. Center, 57905 866 Rd, Concord, NE 68728, USA
e-mail: adatta2@unl.edu; sknezevic2@unl.edu

M.C. Siemens
Department of Agricultural and Biosystems Engineering, University of Arizona, Yuma Agricultural Center, 6425 W 8th St, Yuma, AZ 85364, USA
e-mail: siemens@cals.arizona.edu

S.L. Young and F.J. Pierce (eds.), *Automation: The Future of Weed Control in Cropping Systems*, DOI 10.1007/978-94-007-7512-1_9,

The most likely immediate role for weed control automation in agronomic crops would be in the area of sprayer control so that weed-infested patches could be treated selectively and weed-free patches not treated. If automated weed control solutions are developed and made cost effective and time efficient, then broader adoption of automated weed control technology is possible.

1 Introduction

The needs for and potential roles of automation in agronomic and horticultural crops in the Western Hemisphere are as varied as the crops themselves. Currently, many agronomic systems are based on the use of transgenic, genetically modified, herbicide-tolerant crops that rely, heavily, on post-emergence herbicides, especially glyphosate, for weed management. While the development and extensive adoption of glyphosate-tolerant cotton, field corn, and soybean has provided growers with a short-term, efficient, and cost-effective weed management system, the repeated and exclusive use of glyphosate has resulted in the development of weed communities that are dominated by glyphosate-resistant and glyphosate-tolerant species (Rifai et al. 2002; Geier et al. 2006; Knezevic et al. 2009a, b; Price et al. 2011). A shift toward glyphosate-insensitive weeds suggests that the long-term sustainability of these cropping systems may not be successfully maintained without the appropriation of integrated management programs that limit future herbicide selection pressure (Price et al. 2011; Webster and Sosnoskie 2010). Horticultural crops utilize integrated weed management systems, although the lack of effective herbicides necessitates the use of hand weeding, an endeavor that grows financially less tenable with the rising costs of labor. Automation may be a way to effectively introduce a multi-component, integrated weed management system into agronomic production and to minimize the considerable work force requirements in horticultural crops. Although there has been some adoption of machine-vision guidance systems for cultivators by vegetable growers, the role of automation in North and South American cropping systems has been limited. The potential for an expanded role of automated weed removal is further explored in this chapter.

2 Agronomic Crops: Current Weed Management Practices

Changes in the number, diversity, and timing of herbicide applications in canola, corn, cotton, and soybean have occurred in response to the adoption of glyphosate-tolerant cultivars in North America (Givens et al. 2009a, b; Price et al. 2011; Shaner 2000; Young 2006; USDA-NASS 2012; Webster and Nichols 2012). The most significant herbicide shifts have included reductions in the use of soil-applied, residual herbicides in favor of post-emergence, over-the-top applications of glyphosate for weed control (Givens et al. 2009a, b; Shaner 2000; Young 2006). Despite a diversity of available herbicides, relative to many specialty crops, many producers of agronomic commodities continue to rely on a limited number of mechanisms of action to prevent crop-weed competition and yield loss (Givens et al. 2009b). The amount of acreage planted to glufosinate-tolerant cultivars has increased, in recent years (most notably, in cotton), in response to the subsequent development of glyphosate-resistant Palmer amaranth (*Amaranthus palmeri*). As a consequence, so too has the use of glufosinate for post-emergence weed control. The coming wave of crop technologies for corn, cotton, and soybean include "stacked" cultivars that will express multiple herbicide tolerances to combinations of glyphosate, glufosinate, the auxinic herbicides, ALS inhibitors, acetyl CoA carboxylase (ACCase) inhibitors, and 4-hydroxyphenol pyruvate dioxygenase (HPPD) inhibitors (Green 2012).

In contrast to canola, corn, cotton, and soybean, there is no transgenic glyphosate-tolerant crop-weed control system available for wheat. Weed management programs are still dependent on auxinic herbicides (such as 2,4-D, MCPA, and dicamba) and ALS inhibitors for the control of broadleaf species and ACCase and ALS inhibitors for the management of grasses (Canevari et al. 2007). Additionally, there are conventionally bred, herbicide-resistant wheat varieties (Clearfield™) that are tolerant of imazamox, an ALS inhibitor, which is used to control many key weeds including jointed goat grass (*Aegilops cylindrica* Host), Italian ryegrass (*Lolium multiflorum* Lam.), and some brome (*Bromus*) species (Johnson et al. 2002). Similar to wheat, transgenic glyphosate-tolerant rice is not available. Primary herbicide programs use glyphosate or paraquat for preplant burn down followed by the application of various residual or post-emergence herbicides such as thiobencarb, quinclorac, carfentrazone, propanil, bentazon, or triclopyr. Conventionally, bred rice resistant to imazethapyr and imazamox is currently available for red rice control (Steele et al. 2002).

Pre-plant tillage and in-crop cultivation have proven to be effective tools for controlling emerged weeds and burying their seed below optimal emergence (Anderson et al. 1998; Barberi and Lo Cascio 2001; Feldman et al. 1997). Despite its effectiveness, the use of mechanical weed control in corn, cotton, and soybean has decreased, significantly, over time (Givens et al. 2009a). Conservation tillage practices (including ridge-, mulch-, and no-tillage) have, in many instances, supplanted conventional tillage systems throughout North America. Although reductions in soil disturbance have been associated with improved soil structure and composition, increases in weed seed density have also been reported (Anderson et al. 1998; Barberi and Lo Cascio 2001; Feldman et al. 1997). Cultivators that can be used in-crop (e.g., rotary hoes, flex-tine weeders and spike-tooth harrows, sweep shovels) are nonselective tools; consequently, their use requires significant operator attention and skill to avoid damage to the crop. Furthermore, many growers prefer not to cultivate because the practice is generally slow, as compared to broadcast herbicide applications, and because cultivation activities can disperse weed propagules within and among fields.

Cultural practices, such as the use of crop rotation and cover crops, can be used to maintain weed populations at manageable levels (Price et al. 2011). Crop rotation, independent of herbicides, can alter weed communities via changes to the structure of soil microbial communities and because of differences in crop height, density, and canopy architecture that favor the germination and development of some weed species over others (Leroux et al. 1996; Liebman and Dyck 1993). Cover crops can suppress weeds by serving as a physical barrier to emergence, by inhibiting germination via reduced light transmittance and allelopathic effects, and by preventing herbicide loss through runoff and leaching (Alletto et al. 2010; Krutz et al. 2009; Locke et al. 2005; Price et al. 2006; Reeves et al. 2005).

Weeds are one of the most significant problems in organic crop production systems; effective control requires the use of multiple strategies to achieve economically acceptable weed control and crop yields (Gianessi and Reigner 2007; Hiltbrunner et al. 2007; Place et al. 2009; Stopes and Millington 1991; Wszelaki et al. 2007). Conventional cultivators such as rotary hoes can be used in soybean for use in stale beds and post-plant weed control. Organic soybean producers in the southeastern USA currently execute three to five post-plant rotary hoe operations in soybean (Place et al. 2009). Although multiple rotary hoe operations provide good weed control, soybean yields are often reduced. Conventional cultivator thus may need integration with tools such as flaming or use of an organic herbicide. Midwest organic corn and soybean farmers typically use two to three cultivations during the cropping season (Delate 2011).

3 Agronomic Crops: Opportunities and Challenges

Grower acceptance of glyphosate-tolerant crop technology, especially in the Western Hemisphere, has been driven by improved weed control (including difficult-to-control flora, such as perennial weeds and volunteer crop plants) and reduced crop injury, both of which can lead to higher crop productivity, a reduction in total herbicide inputs, decreased weed management costs, and increased harvest efficiency (Duke et al. 2002; Kuiper et al. 2000; Marshall 1998; Radosevich et al. 1992; Riches and Valverde 2002). Unfortunately, the repeated use of and over-reliance on a single mechanism of action has resulted in the selection for glyphosate-resistant weeds, shifts in weed communities toward glyphosate-tolerant and glyphosate-resistant species, a return to conventional tillage in systems that are dominated by glyphosate-resistant biotypes, and increased costs associated with the adoption of alternate chemical control measures (Rifai et al. 2002; Geier et al. 2006; Knezevic et al. 2009a, b, c; Price et al. 2011).

1. There is need for fast and accurate weed identification in patches so that custom spray mixes can be made on the go automatically.
2. A knowledgeable and well-trained workforce is needed if automation technologies are to be used effectively.
3. Given the level of herbicide-resistant weeds, there is need for research to develop technologies for controlling weeds with tillage or with alternative methods (flame, laser, infrared radiation, electrical energy, and others).

4 Horticultural Crops: Current Weed Management Practices

The contrast between weed management systems in agronomic crops and horticultural crops is distinct. Most flower and vegetable crops use high-intensity tillage in contrast to agronomic crops like corn and soybean which use conservation tillage systems (CTIC 2012). Unlike agronomic crops, which use modern and highly efficient herbicide systems, most herbicides used in vegetable crops are older chemistries that were first released in the 1960s and 1970s when product development and registration costs were much less than today (Fennimore and Doohan 2008). Additionally, many of the vegetable herbicides control a limited number of weeds (e.g., bensulide a major lettuce herbicide, only controls about five broadleaf weed species) (CDMS 2012). Because weed control with herbicides is often incomplete, vegetable growers depend on mechanical cultivation and hand weeding. Integrated weed management is common and necessary in vegetable crops where crop rotation, prevention of new seed introduction, stale seedbed preparation, mechanical cultivation, and hand weeding are commonly used (Lanini et al. 2011). Hand weeding costs in crops like romaine lettuce can exceed $300 per ha (Tourte et al. 2009). Because vegetable growers rely on hand weeding, the most expensive weed control tool, there is considerably more interest in new tools that can reduce the dependency on human labor.

The vegetable herbicide market is small, as compared to agronomic crops for numerous reasons. High crop value, small hectarage per crop, and limited herbicide options could result in a higher potential for liability for herbicide registrants, leading to little financial incentive for herbicide manufacturers to target these commodities with respect to product development (Fennimore and Doohan 2008). For example, although oxyfluorfen can significantly reduce hand weeding costs in onion, its manufacturers hesitate to support IR-4-sponsored food residue studies, because of concerns over potential liability. This situation appears to be getting worse as registration costs increase and herbicide manufactures become more risk averse. Most vegetable herbicides are older products whose maintenance in the market place requires significant funding and effort. One example is the case of pronamide in leaf lettuce, where the food tolerance was lost and the residue package needed to be developed. Additionally, many vegetable crops are small seeded and remain in the seedling stage for longer periods due to their slow growing nature; this makes them more susceptible to stress from herbicides used in the current or preceding crops (Fennimore and Doohan 2008). Even if a herbicide is registered for use, it may not provide complete weed control in the field and will need to be integrated with other tools for growers to achieve acceptable weed control.

Many vegetable producers have also depended on preplant soil fumigants for weed control. The primary fumigant used for decades, methyl bromide (MB), has been classified as a Class I stratospheric ozone-depleting chemical, under the Montreal Protocol; since 2005, use of this fumigant in most of North America is allowed only under critical use exemptions (Anbar et al. 1996; U.S. Environmental

Protection Agency 1993). Most of the Western Hemisphere, other than Canada and the USA, fall under the MB phase out for developing countries which is 2015. The loss of MB has created a void for control of nutsedge in the Southeastern USA, as well as tropical and subtropical America. Crops impacted by the loss of MB are primarily fresh market tomato, strawberries, peppers, eggplant, and cucurbits. The search for alternatives to MB currently includes other synthetic fumigants, bio-fumigants, herbicides, solarization, thermal weed control, and evaluations of various plastic films to improve weed control efficacy.

There are no commercial glyphosate-tolerant vegetable crops, although several transformation events, including one in lettuce, have been field tested (Fennimore and Umeda 2003). Real and perceived issues with market acceptance of transgenic vegetable crops are an important barrier to this technology in these crops. Market acceptance of glyphosate-tolerant feed grains, such as soybean and corn, has been less of a barrier as the buyers are other farmers who are comfortable with transgenic crops. However, for food crops like wheat and vegetables, market acceptance is still lagging (Alston 2004). Produce buyers, processors, and distributors fear a pushback by consumers as crops like lettuce and broccoli are intended for human consumption (i.e., not used for industrial purposes) and consumers do not see any clear benefits of consuming genetically modified vegetables. The diversity of crops under production also slows the adoption of transgenic technology. For example, as many as 60 distinct cultivars of iceberg lettuce alone may be grown throughout the year, and each transgenic event that is carried needs separate evaluation and approval. This makes the research and development process of glyphosate-tolerant vegetable crops expensive and cumbersome (Clark et al. 2004).

Vegetable crops almost always require mechanical cultivation and hand weeding or the use of plastic mulches to achieve weed control. Effective weed control using cultivators depends on numerous factors such as soil type, soil moisture, timing of the operation, and the critical weed-free period for the crop. In relation to the crop plants, Blackmore et al. (2007) defined three weeding zones: interrow, intra-row, and close to crop. Traditional mechanical cultivators are mostly restricted to the interrow zone between crop rows, with some limited intra-row capability and are widely used in vegetable crops. A few examples of available cultivating tools include the flex-tine cultivator, finger weeder, basket weeder, rod weeder, chisel plow, torsion Bezzerides cultivator, Buddingh in-row finger weeder, and brush hoe to remove weeds (Gaskell et al. 2000; Smith et al. 2000). Brief descriptions of many of these implements are available in Cloutier et al. (2007) and SAN (2002). Although these cultivators can readily remove weeds in the interrow zone, the intra-row weeds are more difficult to control and are usually managed by herbicides or hand weeding.

Successful use of cultivators in vegetable cropping systems is an acquired skill that is best described as a combination of science and art (SAN 2002). Controlling weeds with cultivation involves consideration of biological factors like crop and weed growth stage, the selection of the appropriate implement, and correct depth, spacing, and alignment adjustments. When used incorrectly, cultivation tools can damage crop plants, which, occasionally, can result in greater losses than the weed competition would have caused. For example, sweet corn subjected, on one occasion, to blind cultivation with a rotary hoe (between the preemergence herbicide application and the six-leaf stage) did not suffer yield reduction (Leblanc et al. 2006). However, in the same study, three or four sequential cultivations reduced ear number and delayed crop maturity. When used alone, most cultivators do not completely eliminate weeds and are most effective when integrated with other tools such as hand weeding and herbicides. Growers often combine more than one cultivation operation to achieve better weed control. In snap bean, a flex-tine cultivator used alone did not provide acceptable weed control. However, when preceded by a run of a brush hoe or shovel cultivator to achieve weed control, yield was comparable to a broadcast herbicide application (Colquhoun et al. 1999).

Weed control challenges and costs are higher in organic vegetable farms than in conventional farms due to the increased number of cultivations required in one cropping cycle, lack of herbicide options, and an even greater reliance on hand weeding. Weeding in organic vegetables is a costly task second only to the cost of harvesting (Tourte et al. 2009). Although organic growers use the same cultivation tools as conventional growers, organic vegetable production systems are even more reliant on mechanical weed control techniques. More intense use of cultivating and hand weeding often results in even greater weed management costs for organic producers. Weeding and thinning costs in California organic lettuce were estimated at \$941/ha compared to \$445/ha in conventional lettuce (Tourte and Smith 2010; Tourte et al. 2009). In a survey study done in Florida, nearly 90 % of the organic growers indicated that they use some combination of hand hoeing and mechanical cultivation to reduce weed populations (Swisher and Monaghan 1995). Besides cultivation, other weed management strategies used by growers in North America include vegetable transplants, close crop cultivation, flaming of weeds pre-emergence, crop rotation, utilizing cover crops, altering time of nutrient applications, employing the use of plastic mulches, managing the irrigation practices, and soil solarization (Gaskell et al. 2000). While there are no special efforts made in the area of automation for organic farms alone, the organic industry can certainly benefit from advancements in automation to aid in weeding operations.

Direct seeding of vegetables often requires seeding at high population densities and hand thinning the crop to a desired final stand after emergence; incidental weeds can also be removed during this operation (Haar and Fennimore 2003). In lettuce, thinning costs are about \$250 per hectare (Tourte and Smith 2010). In Arizona and California, there is considerable interest in the development of an automatic lettuce thinner to replace hand thinning and reduce the cost of lettuce production. One technology developed and tested for lettuce thinning and weeding operations utilizes a digital camera to detect the location of crop plants and a herbicidal spray to kill unwanted plants (crop or weeds) in a 10 cm wide band (Fig. 9.1). The developers estimate that automatic thinning could reduce labor costs by >25 % in lettuce production (M. Siemens, unpublished data). If lettuce thinning costs could be reduced by >25 %, Arizona and California lettuce growers could save over \$5.8 million annually plus the benefits provided by intra-row weed removal. Several companies are in various stages of commercializing automated thinning machines for lettuce utilizing a similar approach including Agmechtronix, Silver City, NM; Ramsay Highlander, Gonzales, CA; Blue River Technologies, Sunnyvale, CA; and Vision Robotics, San Diego, CA.

Fig. 9.1 Iceberg lettuce thinned by an herbicidal based automated thinning machine (*foreground*), un-thinned lettuce (*background*) (From Siemens et al. 2012)

Cultivators, such as the Robocrop InRow (Garford 2012), were developed in England for use as intra-row cultivators in transplanted vegetable crops such as cabbage and celery (O'Dogherty et al. 2007; Tillett et al. 2008). Robocrop uses a variable-speed rotating, semicircle-shaped disc blade. A camera and computer control guidance system adjusts the rotational speed of the disc in real time so that the opening in the disc blade passes around the transplanted crop. The rotation of the disc allows protection of the crop with the alignment of the opening with the crop plant location and with disc rotation to allow the cutting edge of the disc to remove weeds between the crop plants. The Robocrop InRow cultivator has been evaluated for crop thinning and weed removal in a commercial direct-seeded lettuce planting in California in which lettuce was machine thinned and weeded with the InRow cultivator (Fig. 9.2). After machine thinning with the rotating cultivator and interrow cultivation with the standard cultivator, which did not thin the lettuce, the trial was all thinned and weeded by a hand crew using conventional methods. The standard cultivator treatment had 30–54 % higher weed densities after cultivation than the rotating cultivator; the standard cultivator required 16–31 % more time to thin and weed lettuce than the rotating cultivator (Table 9.1). The goal of automatic lettuce thinning is relevant to weed control in that plant removal in the intra-row space is where both weeds and extra lettuce plants are located. Therefore, the assumption is that devices designed for crop thinning can also be used for intra-row weeding.

Fig. 9.2 Intra-row weeding in celery with the Robocrop InRow cultivator (From Fennimore et al. in press)

Table 9.1 Effects of cultivator type, rotating or standard, on final weed densities and time for a laborer to thin and weed seeded lettuce (Fennimore et al. 2013)

	Trial 1	Trial 4	Trial 5	Trial 1	Trial 4	Trial 5
Cultivator	Weed densities (no. 100^{-2})			Time (sec 100 m^{-1})		
Rotating	1,628.8 b	469.0 b	1118.4 b	649.2 b	474.5 b	411.3 b
Standard	3,466.7 a	912.2 a	1,599.0 a	936.0 a	563.2 a	638.9 a

Means with the same letter within columns for a cultivator or herbicide are not significantly different according to Fisher's protected LSD at $p \leq 0.05$

5 Horticultural Crops: Opportunities and Challenges

Given the costs of developing new herbicides, it is unlikely that more than a few new herbicide products will become available for vegetable crops in the foreseeable future. The limited acreage of many vegetable crops means that the market value for herbicides used in vegetable is much less than for agronomic crops and chemical companies focus their research on the large markets (Gast 2008). Similarly, large machinery manufacturers tend to focus on agronomic crops where the largest sales potential exists, rather than on relatively small acreage vegetable crops. Government-sponsored programs, like the USDA's IR-4 project and Agri-Food Canada's minor use program, facilitate the registration of herbicides and other pesticides on specialty crops to ease the concerns about the lack of economic incentives (Kunkel et al. 2008). Perhaps a program similar to IR-4 is needed to facilitate research on and development of technology for weed removal automation in vegetables and other horticultural crops. Such a program may provide a means to incentivize manufacturers to produce machinery for specialty crops. Because weed control problems in vegetable crops are not likely to be solved with new herbicide registrations or herbicide-resistant cultivars, automation technology holds greater promise for both conventional and organic horticultural cropping systems.

In the USA, labor issues and immigration policy reforms are a matter of constant discussion and debate with important implications to weed control. There is a great demand for labor in vegetables, and hand weeding consumes a significant percentage of the labor budget for these crops. In 1999–2000, 55 % of the laborers hired in the USA were not authorized to work legally (Levine 2007). While there is no substantial evidence yet, it is likely that automation of weed control would reduce the need for manual labor. In addition to alleviating labor shortage problems in many geographic areas of North America, more efficient weed control due to automation could reduce labor scheduling issues due to time conflicts among weed control and other farm activities, such as harvesting in a complex vegetable production system.

The lack of personnel with sufficient training and background to operate automated machinery is likely to become an issue. Traditional farm workers are, generally, poorly educated and not computer literate and, thus, are poor candidates to operate automatic technology. The operation of automatic equipment creates the need for a better trained work force than has been required for traditional cultivator operations in the past. The challenge is to justify the cost of automatic equipment and trained personnel with increased efficiency and commensurate production cost reductions that pay for the high machinery costs.

6 Orchard and Vineyard Crops: Current Weed Management Practices

Orchard and vineyard crops in the Western Hemisphere include a wide variety of fruiting trees, vines, and shrubs. This diverse group of perennial crops includes nut crops (e.g., almond, walnut, pistachio, cashew, macadamia), stone fruit (e.g., peach, plum, cherry, apricot, nectarine), pome fruit (e.g., apple, pear, quince), citrus (e.g., orange, lemon, grapefruit), grape (wine, table, and raisin types), and berries (e.g., blueberry, raspberry, blackberry) as well as a number of tropical and subtropical species including cacao, coconut, coffee, mango, papaya, pineapple, and avocado among others. The most economically important of these are typically grown in intense monoculture plantings with a lifespan ranging from multiple years to several decades or more. Weed management is a major factor affecting the establishment, production, and harvest in orchards and vineyards; however, weed control practices can vary tremendously among specific crops, between production regions, and even during the lifespan of an individual planting.

Large-statured tree and vine crops are very competitive with weeds once a canopy is established; however, wide intra- and interrow spacing and slow initial growth can leave them very vulnerable to weed competition in the first few years after planting. Young trees and vines have relatively small root systems, and keeping a 0.6–1.2 m weed-free radius around the trunk during the early establishment stages is needed to reduce competition for soil resources. Additionally, effective weed control minimizes competition for light, especially in summer when weeds can grow higher than the young trees. Good weed management during the long period of orchard establishment will hasten nonbearing trees into production and promote the long-term health and productivity of the crop.

Excessive weed populations can also reduce cropping system efficiency in established orchards and vineyards and can increase problems with other pests (Ashigh and Marquez 2010). For example, weeds can serve as alternate hosts or provide habitat for insect, mite, disease, and vertebrate pests. Weeds can interfere with irrigation efficiency by physically blocking irrigation furrows or by encroaching into or altering the patterns of low-volume (drip and fan-jet) irrigation emitters. Applications of pesticides, growth hormones, and foliar and soil nutrients are more efficient where weeds are managed. Similarly, cultural operations (e.g., irrigation, thinning, pruning, training, harvesting) are easier when weeds are not present to slow or hinder workers or machinery.

Orchard and vineyard weeds are managed in many different ways throughout the Western Hemisphere depending on climate, soil, weed pressure, crop age, planting arrangement, and irrigation systems. Growers typically use a combination of weed control tactics that best fit their specific production system and limitations on available time, equipment, and economic resources. In contrast to many other cropping systems, weed management strategies may differ both within and between crop rows. Because of spatial and temporal differences relative to annual agronomic and horticultural crops, tree and vine cropping systems present unique opportunities and challenges for automated weed control.

Consumer interest in organic and locally produced foods as subsequent market premiums has led to a substantial increase in organic orchard and vineyard production systems in the Western Hemisphere, a trend that is strongest in North America. Although still less than 4 % of the total acreage in the USA, organic acreage of tree nuts, citrus, apples, grapes, berries, and subtropical fruits increased two- to fivefold between 2000 and 2008 (USDA-ERS). The largest individual organic orchard and vineyard crops include over 12,000 ha of grapes, 7,700 ha of apples, 10,000 ha of tree nuts, and 6,200 ha of citrus.

Weed control in organic orchards and vineyards depends heavily on mechanical tactics such as tillage, mowing, and flaming (Granatstein 2012). Additionally, some growers suppress or manage weed populations with cover crop management, hand labor (hoeing, pulling, and string trimmers), synthetic or organic mulches, and organic herbicides. Although weed competition can be extremely detrimental to organic agronomic and vegetable crops, once organic orchards and vineyards are established, weeds usually are not the most limiting pest management issues. However, in both young and established organic orchards and vineyards, weed management tactics represent substantial upfront or ongoing costs and are a significant economic concern.

Herbicides are commonly used in orchard and vineyard production systems in the Western Hemisphere. A number of preemergence and postemergence herbicides are registered in orchard and vineyard crops, but specific product choices vary depending on the specific crop and country. Herbicide options also can be different in new plantings (nonbearing) compared to established (bearing) orchards or vineyards because of crop safety concerns for the young plants.

Herbicides are used across the orchard or vineyard floor in some situations; however, they are more commonly applied in intra-row strips centered on the crop row, while weeds in the interrow area are controlled using nonchemical tactics such as mowing or tillage. Herbicide selectivity in trees and vines is often achieved by selecting an appropriate application timing (such as the winter dormant season) and directed spray applications that minimize crop exposure, by using shielded sprayers or trunk guards, or by otherwise exploiting the differences between the crop and weed rooting depth or canopy height in order to expose only the weed to the herbicide. Because of opportunities for herbicidal weed control with broad-spectrum herbicides, many orchard and vineyard systems have become heavily dependent on postemergence herbicides, especially glyphosate and paraquat due to their relatively low cost. For example, apple, grape, and citrus orchards weed control in Argentina, Brazil, and Chile is achieved through several applications of glyphosate in a year (Vila-Aiub et al. 2008). Similarly, glyphosate is the most widely used herbicide in every orchard and vineyard crop in California (CDPR 2012). Similar to agronomic crops, the reliance on a single mode of action herbicide has led to selection of glyphosate-resistant weeds or shifts to tolerant species in several orchard and vineyard cropping systems.

A wide array of mechanical cultivation equipment can be used to control weeds in orchard and vineyard production systems. The choice of cultivator is determined by crop biology and planting scheme (interrow and intra-row spacing) as well as the availability of the machinery. In both young and established orchards, frequent and shallow cultivation with light tillage implements such as discs, spring tooth harrows, and orchard knives can be used to uproot small weeds and dry surface soils to reduce weed establishment. In young plantings, tillage equipment can be operated quite close to the crop row; however, care must be taken to avoid physical damage to the trunks, crowns, bark tissue, and shallow roots of the tree or vine. Excessive tillage can have detrimental effects on soil structure, crop productivity, and weed control. Tillage can bring buried weed seed close to the surface where they can germinate and spread vegetative portions of perennial weeds around the field (Ashigh and Marquez 2010). Continuous tillage degrades soil structure and can create a plow layer that retards root and water penetration. Soil erosion can be a problem if tillage is used in orchards and vineyards on hilly terrain. Some perennial crops such as avocado and citrus have relatively shallow roots, and frequent tillage can prune these roots and restrict the volume of soil the plant is able to explore.

The relatively wide spacing of many orchard crops permits intra-row weed control using cross tillage (tillage in two perpendicular directions); however, irrigation systems and high-density plantings can limit this technique. Within-row cultivation can also be accomplished in orchards and vineyards using implements such as the French plow and Bezzerides cultivator with a mechanical trigger to move the equipment around the established crop. Within-row cultivation is limited due to concerns about excessive tree damage, which can include root damage from cultivation blades or trunk damage caused by the trigger system (Shrestha et al. 2013). In many cases, intra-row cultivation is also limited by slow operational speeds (Khot et al. 2006) and by the same issues affecting interrow cultivation, such as soil structure degradation, dust, and spreading of perennial weed propagules.

Mowing is commonly used for weed management between crop rows in established orchards and vineyards and is commonly combined with strip herbicide applications. Mowing is relatively inexpensive, fast, and controls weeds outside of the weed-free strip. During crop establishment, it may be necessary to mow up to 10 times per year to keep weeds under control; however, fewer mowing operations are typically needed in established crops. Regional precipitation and irrigation practices influence the frequency of mowing. In arid regions, fewer summer mowing operations are typically needed in crops using drip irrigation as compared to sprinkler irrigation because the nonirrigated areas do not support significant weed growth during the dry part of the growing season. Excessive field operations (including mowing) during the early spring can lead to soil compaction problems or the development of ruts that may affect other field operations.

Recent automated weed control work in orchards and vineyards has evaluated sensor technology to identify and selectively treat weeds with a herbicide or other control treatment. For example, Stover et al. (2003) and Rector (2007) used sensor-activated spray systems in orchards to apply herbicide only to areas with weed canopies which reduced herbicide use by 25–36 % in lemon and pecan, respectively. One such spray sensor system commercially available to growers in the North American market is the WeedSeeker by Trimble Navigation Ltd. This system can significantly reduce post-emergence herbicide use by applying the spray solution only where weeds are present; however, the high cost of the technology combined with the relatively low cost of broad-spectrum herbicides currently limits adoption (McGinley 2001). Khot et al. (2006) suggest sensor fusion, for example, combining a GPS sensor with dynamic measurement unit sensor for better navigational capabilities, which could prove useful in tasks such as intra-row weeding in nursery tree plantations.

The large trunks of tree and vine crops (or the trellis supports used in some crops) could offer a means of real-time identification of crop plants in orchard or vineyards that would not depend on GPS. A relatively simple sensor could detect crop plants and direct subsequent intra-row physical, chemical, or thermal weed control operations. In trials conducted in Pennsylvania and Washington, Hamner et al. (2010) describe the evaluation of a vehicle that uses laser range scanners to detect trees and other objects in its vicinity, builds a model of the row of trees, and uses this model to steer the vehicle along the row without GPS. This system has the capability to determine when the tree row ends or to spot missing trees within a row. Besides the laser scanner, other necessary sensors include the wheel encoders that measure the distance traveled and the steering angle. Such a system can be used for several routine tasks in the orchard that may allow for mechanization including harvesting, pre-thinning, measuring tree diameter, and mowing.

7 Orchard and Vineyard Crops: Opportunities and Challenges

Orchard and vineyard crops in the Western Hemisphere present several opportunities and challenges to the development and widespread adoption of automated weed control strategies. Although opportunities vary considerably among specific crops and countries due to production intensity, crop value, available capital, and labor costs, several attributes make this system attractive for development and implementation of automated and precision weed control techniques.

First, once established, a substantial size difference between the crop and weed seedlings presents opportunities for identification and selective treatment of the weeds. Additionally, established woody trees and vines are generally more tolerant of chemical and physical control tactics compared to newly emerged weeds which reduce the required specificity of any control measure. Second, the long lifespan of these perennial crops and unchanging planting arrangements make it possible to map the coordinates of the crop plants and utilize those data for numerous operations, including weed control, over several years. Semiautomated "smart" sprayers or cultivators can utilize the crop location data layer in a GIS to replace, simplify, or speed up the plant identification computations.

The relatively high value of orchard and vineyard crops, plus the increasing interest in organically produced foods, may offer another economic opportunity for automated weed control in these crops. Some organic weed control tactics such as flaming and organic herbicides are not always economically viable using current technologies or sufficiently safe to young crop plants. However, if these weed management tools could be used more selectively (e.g., only when weeds were present) using improved plant sensor technology, they may become more acceptable and adoptable by orchard and vineyard operators.

Finally, the regular grid- or row-planting schemes common to orchard and vineyards may allow one of the most promising opportunities for truly automated weed control that simultaneously integrates sensor technologies to identify crops and weeds, selective or nonselective control strategies, and unmanned machines to navigate through the crop rows (Griepentrog et al. 2009; Zijlstra et al. 2011).

Some of the challenges to automated weed control in orchards and vineyards are similar to other cropping systems, while others are more specific. Technology costs, especially for early adopters, often can be prohibitive and be a significant barrier to adoption for smaller, less well-capitalized operations. In particular, many organic growers tend to be smaller-scale operations with more limited budgets. Although they could benefit most from new or improved nonchemical weed control methods, they may not be the first to adopt new technologies unless if fits other crops in the organic production system. Another barrier to adoption is the required level of expertise required to set up, operate, and troubleshoot a highly technical system that may integrate a variety of electronic, mechanical, and computer components as well as the more traditional weed control mechanisms. Absent a clear production or economic incentive, adoption of expensive and complicated automated weed control equipment is likely to proceed at a moderate rate. However, development and adoption rates could change drastically if other weed control techniques become unavailable, unsuccessful, or cost prohibitive.

Other challenges to utilization of robotic and automated weed control in orchards and vineyards include developing robust systems that can operate successfully in tight spaces and around irrigation systems, detect and avoid unexpected objects (workers, branches, harvest equipment, etc.), and operate under dense crop canopies that can hinder GPS visibility. Finally, automated weed control equipment must be able to provide acceptable weed control without damaging the crop (roots, limbs, or fruits) to a greater degree than conventional techniques.

8 Conclusions

Regardless of the actual technique used to manage weeds, advancements in sensor and geospatial technology will allow opportunities for selective and precise use of existing weed control strategies in an automated or semi-automated way in cropping systems of the Western Hemisphere. As market preferences, environmental concerns, regulation, and labor availability change, increasingly automated weed control may allow growers to continue to economically produce safe and high-quality vegetables, grains, fruits, and nuts.

References

Alletto L, Coquet Y, Benoit D, Heddadj D, Barriuso E (2010) Tillage management effects on pesticide fate in soils. A review. Agron Sustain Dev 30:367–400

Alston JM (2004) Horticultural biotechnology faces significant economic and market barriers. Calif Agric 58:80–88

Anbar AD, Young YL, Chavez FP (1996) Methyl bromide: ocean sources, ocean sinks, and climate sensitivity. Global Biogeochem Cycles 10:175–190

Anderson RL, Tanaka DL, Black AL, Schweizer EE (1998) Weed community and species response to crop rotation, tillage and nitrogen fertility. Weed Technol 12:531–536

Ashigh J, Marquez EE (2010) Integrated weed management in pecan orchards. New Mexico State University Cooperative Extension Publication, Guide H-656. Available via http://pecan.ipmpipe.org/library/content/intgd_weed_mgmt_h656.pdf. Accessed 7 Jan 2012

Barberi P, Lo Cascio B (2001) Long-term tillage and crop rotation effects on weed seedbank size and composition. Weed Res 41:325–340

Blackmore BS, Griepentrog HW, Fountas S, Gemtos TA (2007) A specification for an autonomous crop production mechanization system. Agricultural Engineering International: the CIGR Ejournal. Manuscript PM 06 032, vol IX, Sept 2007

California Department of Pesticide Regulation [CDPR] (2012) Summary of pesticide use report data. Available at http://www.cdpr.ca.gov/docs/pur/pur10rep/comrpt10.pdf. Accessed 16 Apr 2013

Canevari WM, Wright SD, Carlson HL, Jackson LF (2007) Small grains integrated weed management. University of California, Division of Agriculture and Natural Resources Publication 3466. Available at http://www.ipm.ucdavis.edu/PMG/r730700111.html. Accessed 18 Apr 2013

Clark D, Klee H, Dandekar A (2004) Despite benefits, commercialization of transgenic horticultural crops lags. Calif Agric 58:89–98

Cloutier DC, van der Weide RY, Peruzzi A, LeBlanc ML (2007) Mechanical weed management. In: Upadhyaya MK, Blackshaw RE (eds) Non-chemical weed management. CAB International, Oxfordshire, pp 111–134

Colquhoun JB, Bellinder RR, Kirkwyland JJ (1999) Efficacy of mechanical cultivation with and without herbicides in broccoli (*Brassica oleracea*), snap bean (*Phaseolus vulgaris*) and sweet corn (*Zea mays*). Weed Technol 13:244–252

Conservation Technology Information Center [CTIC] (2012) National crop residue management survey. Available at http://www.ctic.purdue.edu/. Accessed 17 Apr 2013

Crop Data Management System [CDMS] (2012) Prefar: specimen label. Gowan Company. Available at http://www.cdms.net/LDat/ld14201l.pdf. Accessed 27 July 2012

Delate K (2011) Weed management for organic farmers. Iowa State University, University Extension Publication PM 1883. Available at http://extension.agron.iastate.edu/organicag/pubs/PM1883.pdf. Accessed 28 Dec 2011

Duke SO, Scheffler BE, Dayan FE, Dyer WE (2002) Genetic engineering crops for improved weed management traits. Crop Biotechnol 829:52–66

Feldman SR, Alzugaray C, Torres PS, Lewis P (1997) The effect of different tillage systems on the composition of the seedbank. Weed Res 37:71–76

Fennimore SA, Doohan DJ (2008) The challenges of specialty crop weed control, future directions. Weed Technol 22:364–372

Fennimore SA, Umeda K (2003) Time of glyphosate application in glyphosate-tolerant lettuce. Weed Technol 17:738–746

Fennimore SA, Smith RF, Tourte L, LeStrange M, Rachuy JS (in press) Evaluation and economics of a rotating intra-row cultivator in celery and leafy greens. Weed Technol (in review)

Garford (2012) Robocrop in row weeder. Available at http://www.garford.com/products_robocropinrow.html. Accessed 23 May 2013

Gaskell M, Fouche B, Koike S, Lanini T, Mitchell J, Smith R (2000) Organic vegetable production in California-Science and practice. HortTechnology 10:699–713

Gast RE (2008) Industry views of minor crop weed control. Weed Technol 22:385–388

Geier PW, Stahlman PW, Frihauf JC (2006) KIH-485 and *S*-metolachlor efficacy comparisons in conventional and no-tillage corn. Weed Technol 20:622–626

Gianessi LP, Reigner NP (2007) The value of herbicides in U.S. crop production. Weed Technol 21:559–566

Givens WA, Shaw DR, Kruger GR, Johnson WG, Weller SC, Young BG, Wilson RG, Owen MDK, Jordan D (2009a) Survey of tillage trends following the adoption of glyphosate-resistant crops. Weed Technol 23:150–155

Givens WA, Shaw DR, Kruger GR, Johnson WG, Weller SC, Young BG, Wilson RG, Owen MDK, Jordan D (2009b) A grower survey of herbicide use patterns in glyphosate-resistant cropping systems. Weed Technol 23:156–161

Granatstein D (2012) Tree fruit production with organic farming methods. Available at http://organic.tfrec.wsu.edu/OrganicIFP/OrganicFruitProduction/OrganicMgt.PDF. Accessed 7 Jan 2012

Green JM (2012) The benefits of herbicide-resistant crops. Pest Manag Sci 68:1323–1331

Griepentrog HW, Andersen NA, Andersen JC, Blanke M, Heinemann O, Madsen TE, Nielsen J, Pedersen SM, Ravn O, Wulfsohn D (2009) Safe and reliable: further development of a field robot. In: van Henten EJ, Goense D, Lokhorst C (eds) Precision agriculture '09. Wageningen Academic Publishers, Wageningen, pp 857–866

Haar MJ, Fennimore SA (2003) Evaluation of integrated practices for common purslane management in lettuce. Weed Technol 17:229–233

Hamner B, Singh S, Bergerman M (2010) Improving orchard efficiency with autonomous utility vehicles. ASABE annual international meeting. Paper number 1009415

Hiltbrunner J, Liedgens M, Bloch L, Stamp P, Streit B (2007) Legume cover crops as living mulches for organic wheat: components of biomass and the control of weeds. Eur J Agron 26:21–29

Johnson J, Haley S, Westra P (2002) Clearfield wheat. Colorado State University Cooperative Extension Publication no. 3.116. Available at http://wheat.colostate.edu/CSUWheatBreeding/Links_files/03116.pdf. Accessed 30 Apr 2013

Khot LR, Tang L, Blackmore SB, Nørremark M (2006) Navigational context recognition for an autonomous robot in a simulated tree plantation. Trans ASAE 49:1579–1588

Knezevic SZ, Datta A, Ulloa SM (2009a) Growth stage impacts tolerance to broadcast flaming in agronomic crops. In: Cloutier DC (ed) Proceedings of the 8th European Weed Research Society workshop on physical and cultural weed control, Zaragoza, Spain, pp 86–91

Knezevic SZ, Costa CM, Ulloa SM, Datta A (2009b) Response of maize (*Zea mays* L.) types to broadcast flaming. In: Cloutier DC (ed) Proceedings of the 8th European Weed Research Society workshop on physical and cultural weed control, Zaragoza, Spain, pp 92–97

Knezevic SZ, Datta A, Ulloa SM (2009c) Tolerance of selected weed species to broadcast flaming at different growth stages. In: Cloutier DC (ed) Proceedings of the 8th European Weed Research Society workshop on physical and cultural weed control, Zaragoza, Spain, pp 98–103

Krutz LJ, Locke MA, Steinreide RW Jr (2009) Interactions of tillage and cover crop on water, sediment, and pre-emergence herbicide loss in glyphosate-resistant cotton: implication for control of glyphosate-resistant weed biotypes. J Environ Qual 38:1240–1247

Kuiper HA, Kleter GA, Noordam MY (2000) Risks of the release of transgenic herbicide-resistant plants with respect to humans, animals and the environment. Crop Prot 19:773–778

Kunkel DL, Salzman FP, Arsenovic M, Baron JJ, Braverman MP, Holm RE (2008) The role of IR-4 in the Herbicide registration process for specialty food crops. Weed Technol 22:373–377

Lanini WT, Hembree KJ, Miyao G, Stoddard CS (2011) Weeds. In UC IPM pest management guideline: Tomato. University of California, Division of Agriculture and Natural Resources Publication 3470. Available at http://www.ipm.ucdavis.edu/PMG/r783700111.html. Accessed 18 Apr 2013

Leblanc ML, Cloutier DC, Stewart KA (2006) Rotary hoe cultivation in sweet corn. HortTechnology 16:583–589

Leroux GD, Benoit DL, Banville S (1996) Effect of crop rotations on weed control, Bidens cernua and Erigeron canadensis populations, and carrot yields in organic soils. Crop Prot 15:171–178

Levine L (2007) Farm labor shortages and immigration policy. Congressional Research Service Report for Congress. Available at http://digitalcommons.ilr.cornell.edu/key_workplace/23/. Accessed 27 Dec 2011

Liebman M, Dyck E (1993) Crop rotations and intercropping strategies for weed management. Ecol Appl 3:92–122

Locke MA, Zablotowicz RM, Bauer PJ, Steinreide RW, Gaston LA (2005) Conservation cotton production in the southern United States: herbicide dissipation in soil and cover crops. Weed Sci 53:717–727

Marshall G (1998) Herbicide-tolerant crops – real farmer opportunity or potential environmental problem? Pest Sci 52:394–402

McGinley S (2001) Fighting weeds in lemon orchards. Weed-sensing sprayer could be the key. Available at http://ag.arizona.edu/pubs/general/resrpt2001/10.html. Accessed 12 Jan 2012

O'Dogherty MJ, Godwin RJ, Dedousis AP, Brighton LJ, Tillett ND (2007) A mathematical model of the kinematics of a rotating disc for inter- and intra-row hoeing. Biosyst Eng 96:169–179

Place GT, Reberg-Horton SC, Burton MG (2009) Effects of preplant and postplant rotary hoe use on weed control, soybean pod position, and soybean yield. Weed Sci 57:290–295

Price AL, Reeves DW, Patterson MG (2006) Evaluation of weed control provided by three winter cereals in conservation-tillage soybean. Renew Agric Food Syst 21:159–164

Price AJ, Balkcom KS, Culpepper SA, Kelton JA, Nichols RL, Schomberg H (2011) Glyphosate-resistant Palmer amaranth: a threat to conservation tillage. J Soil Water Conserv 66:265–275

Radosevich SR, Ghersa CM, Comstock G (1992) Concerns a weed scientist might have about herbicide-tolerant crops. Weed Technol 6:635–639

Rector RJ (2007) The use of an automatic spot-sprayer in western tree crops and weed control in a pecan orchard using preemergence and postemergence herbicides. University of Arizona PhD dissertation, 201 p

Reeves DW, Price AJ, Patterson MJ (2005) Evaluation of three winter cereals for weed control in conservation-tillage non-transgenic cotton. Weed Technol 19:731–736

Riches CR, Valverde BE (2002) Agricultural and biological diversity in Latin America: implications for development, testing, and commercialization of herbicide resistant crops. Weed Technol 16:200–214

Rifai MN, Astatkie T, Lacko-Bartosova M, Gadus J (2002) Effect of two different thermal units and three types of mulch on weeds in apple orchards. J Environ Eng Sci 1:331–338

Shaner DL (2000) The impact of glyphosate-tolerant crops on the use of other herbicides and on resistance management. Pest Manag Sci 56:320–326

Shrestha A, KaanKurtural S, Fidelibus MW, Dervishian G, Konduru S (2013) Efficiency and cost-effectiveness of cultivators, steam, or an organic herbicide in weed control and vine productivity in organic vineyards in the San Joaquin Valley. HortTechnology 23:99–108

Siemens MC, Herbon R, Gayler RR, Nolte KD, Brooks D (2012) Automated machine for thinning lettuce – Evaluation and development. ASABE paper no 12-1338169. ASABE, St. Joseph, pp 14

Smith R, Lanini WT, Gaskell M, Mitchell J, Koike ST, Fouche C (2000) Weed management for organic crops. University of California, Division of Agriculture and Natural Resources. Publication 7250

Steele GL, Chandler JM, McCauley GN (2002) Control of red rice (*Oryza sativa*) in imidazolinone-tolerant rice (*O. sativa*). Weed Technol 16:627–630

Stopes C, Millington S (1991) Weed control in organic farming systems. In: Proceedings of the Brighton crop protection conference-weeds, Brighton, UK, pp 185–192

Stover E, Salvatore J, Wirth F (2003) Pesticide spray reduction from using a sensor-actuated spray system in Indian River grapefruit. HortTechnology 13:178–181

Sustainable Agriculture Network-SAN (2002) In: Bowman G (ed) Steel in the field. A farmers guide to weed management tools. Sustainable Agriculture Network, Beltsville

Swisher ME, Monaghan P (1995) Florida's organic farmers: a profile. Available at http://www.joe.org/joe/1995june/rb1.php. Accessed 28 Dec 2011

Tillett ND, Hague T, Grundy AC, Dedousis AP (2008) Mechanical within-row weed control for transplanted crops using computer vision. Biosyst Eng 99:171–178

Tourte L, Smith RF (2010) Sample production costs for wrapped iceberg lettuce sprinkler irrigated – 40-inch Beds. University of California Cooperative Extension Publication LT-CC-10. Available from http://coststudies.ucdavis.edu/files/2010Lettuce_Wrap_CC.pdf. Accessed 23 May 2013

Tourte L, Smith RF, Klonsky KM, De Moura RL (2009) Sample costs to produce organic leaf lettuce. University of California Cooperative Extension Publication LT-CC-09-O. Available at http://coststudies.ucdavis.edu/files/lettuceleaforganiccc09.pdf Accessed 31 Aug 2013

U. S. Environmental Protection Agency (1993) Protection of stratospheric ozone. Fed Regist 58:15014–15049

United States Department of Agriculture National Agricultural Statistics Service [USDA-NASS] (2012) Statistics by subject. Available at http://www.nass.usda.gov/. Accessed 2 July 2012

Vila-Aiub MM, Vidal RA, Balbi MC, Gundel PE, Trucco F, Ghersa CM (2008) Glyphosate-resistant weeds of South American cropping systems: an overview. Pest Manag Sci 64:366–371

Webster TM, Nichols RL (2012) Changes in the prevalence of weed species in the major agronomic crops of the Southern United States: 1994/1995 to 2008/2009. Weed Sci 60:145–157

Webster TM, Sosnoskie LM (2010) The loss of glyphosate efficacy: a changing weed spectrum in Georgia cotton. Weed Sci 5:73–79

Wszelaki AL, Doohan DJ, Alexandrou A (2007) Weed control and crop quality in cabbage [*Brassica oleracea* (capitata group)] and tomato (*Lycopersicon lycopersicum*) using a propane flamer. Crop Prot 26:134–144

Young BG (2006) Changes in herbicide use patterns and production practices resulting from glyphosate-resistant crops. Weed Technol 21:301–307

Zijlstra C, Lund I, Justesen AF, Nicolaisen M, Jensen PK, Bianciotto V, Posta K, Balestrini R, Przetakiewicz A, Czembor E, van de Zande J (2011) Combining novel monitoring tools and precision application technologies for integrated high-tech crop protection in the future. Pest Manag Sci 67:616–625

Chapter 10
Field Applications of Automated Weed Control: Northwest Europe

Jan Willem Hofstee and Ard T. Nieuwenhuizen

Abstract In Northwest Europe there is high need for advanced weed control methods. The use of crop protection chemicals has become stricter, and integrated pest management is required by regulations from the European Union. This need has resulted in the development of several advanced weed control principles based on a combination of proven technologies in combination with decision systems. A major problem with full-field-based methods is that the required settings depend very much on the specific conditions. Use of decision systems helps to improve these methods. Emerging new technologies as machine vision and GPS enabled more precise methods focused on the interrow and intrarow zone and on the plant itself. Some of the methods have already achieved a high level of development and resulted in commercially available weed control equipment with sensors and actuators for precise control. This chapter discusses the advancements achieved in NW Europe on mechanical weed control (full field, interrow and intrarow), physical weed control (steaming and flaming) and chemical weed control (full field, spot and plant oriented).

J.W. Hofstee (✉)
Farm Technology Group, Wageningen University, Radix, Droevendaalsesteeg 1, 6708 PB Wageningen, The Netherlands
e-mail: janwillem.hofstee@wur.nl

A.T. Nieuwenhuizen
Plant Research International, Field Technology Innovations Group, Wageningen UR, Radix, Droevendaalsesteeg 1, 6708 PB Wageningen, The Netherlands
e-mail: ard.nieuwenhuizen@wur.nl

S.L. Young and F.J. Pierce (eds.), *Automation: The Future of Weed Control in Cropping Systems*, DOI 10.1007/978-94-007-7512-1_10,

1 Introduction

Weed control is still a major issue in modern agriculture, despite the development of several tools for weed removal in the past decades. According to Oerke (2006), overall potential losses (i.e. without crop protection) ascribed to weeds are 34 %. Weed management has always been one of the key issues in most agricultural systems (Kropff et al. 2008). The need for precision weed control in the future is increasing, in NW Europe and around the world. In NW Europe, which includes the UK, Ireland, Scandinavian countries, Germany, Belgium, France, the Netherlands and Luxembourg, the policy on the use of crop protection chemicals has become stricter, and in addition the number of allowed chemicals for crop protection is reducing. A parallel development is the increase of labour costs and at the same time a decreasing availability of skilled labour for work in agriculture, including weed control. The intensive cropping systems with high yields per hectare in NW Europe require an effective weed control to realise these high yields. Where in the past the trend for weed control in conventional agriculture was towards chemical weed control, we now see a trend towards more non-chemical weed control also in conventional (non-organic) farming. The strict policies on the use of crop protection chemicals in conjunction with the development of integrated crop protection schemes reinforce this trend towards an integrated pest management approach. This integrated approach is required to comply with societal demands on safe food production and environmental protection. Integrated pest management is also required by regulations from the European Union (European Commission 2009). As part of this integrated approach, cropping systems have to be changed and weed control methods after they have emerged have to be improved. An example of this change in cropping system is the precision drilling of seeds (Griepentrog et al. 2005) leading to new opportunities in crop management. An example is GeoSeed seeding principle by the Kverneland Group enabling hoeing in four different directions.

In NW Europe a wide variety of crops is grown. The most important crops in area are cereals as wheat and barley, sugar beets, potatoes and maize (mainly for silage). Besides that, at several places vegetable crops like onions, carrot, cabbage, lettuce and leek are grown. Other important crops are beans, peas, rape or flower bulbs. In NW Europe no rice, soybeans or cotton is grown. Especially in the smaller crops, often high-value crops, weed control is becoming more and more a problem due to the restriction of the use of crop protection chemicals, and several of these have very open canopy structure in the beginning of the growth, giving ample space to weeds.

Mechanical weed control was for some decades back the domain of mainly organic farmers. Conventional farmers considered mechanical weed control as old fashioned and techniques of the past. The introduction of automation and electronics in agriculture opened new roads for development of modern weed control. Where in the past the organic farmers were the driving force for development in mechanical weed control, we now see that these methods are more and more adopted by conventional farmers as part of their integrated crop protection scheme. This trend also fits

Table 10.1 Overview of methods for weed control, the corresponding machines and crops

Method	Machine	Crops
Mechanical control	Spring tines	Cereals
	Interrow weeding	All crops growing in rows with spacing >= 25 cm
	Intrarow weeding	Transplanted and precision-seeded row crops
Physical control	Steaming	All row crops
	Flaming	All row crops
Chemical control	Full-field broadcast spraying	All crops
	Full-field spot spraying	Cereals
	Plant-specific spraying	Transplanted and precision-seeded row crops

the development of more precise chemical control methods, focused on only applying the chemicals where needed and in quantity related to the need instead of whole-field application of maximum amounts to be sure that all weeds are controlled.

Over the past two decades, several weed sensing systems and precision implements were developed, but there are still some barriers. Two main barriers are the lack of truly robust weed recognition methods and limitations in the capacity of highly accurate spraying and weeding apparatus (Christensen et al. 2009).

In this chapter the main focus is on research on weed control and weed control applications in NW Europe presented in 2005 or later. The research shows the trend towards the future and is the basis for the further development of advanced weed control in NW Europe and the driving force for modern weed control where low or no input of chemicals is desired or even required. Where available, results from practice are presented, especially if they are a spin-off of previous research. Table 10.1 gives for NW Europe an overview of the machinery available for each method and the type of crops where the machinery is applied.

2 Applications: Mechanical Weed Control

Mechanical weed control methods are in general attractive because of the high capacity, wide applicability and low costs (Dedousis and Godwin 2005). A major limitation however is the variable effectiveness and the limited selectivity at early crop stages.

2.1 Full-Field Oriented

There is a wide range of tools available for mechanical weed control, and overviews are given, for example, by Cloutier et al. (2007) and Weide et al. (2008). Cloutier et al. (2007) present an overview of three main techniques for weed management: use of tillage, cutting weeds and pulling weeds. These techniques are very general and in principle applicable worldwide. Weide et al. (2008) describe the state of the art of different machinery for mechanical intrarow weed control. They also discussed the short-comings and prospects for further research, development and implementation of mechanical intrarow weed control. This machinery is usually used in crops grown on rows that have open canopy architecture like sugar beet, cabbages, carrots and onions. There is a large number of studies available on whole crop-oriented weed control (Melander et al. 2005; Kurstjens 2007; Wei et al. 2010).

Important aspects are timing and intensity of the weed control operation. An important criterion is the selectivity of the operation, which is affected by timing and intensity, and, of course, the method itself and also depends on the crop and weed species. The main problem with whole crop-oriented weed control is that the required settings (i.e. aggressiveness of the cultivation) are very situation dependent. Ferrero et al. (2007) studied the mechanical weed control in organic soybeans and maize. Methods used were flame weeding, spring tine harrowing and interrow hoeing and were applied at different growth stages. They conclude that these methods have a promising potential, but they also mention that the intervention timing is a crucial factor to have effective weed control together with crop selectivity. Rasmussen et al. (2008) investigated the effects of row spacing, timing, direction and orientation on crop/weed selectivity on post-emergence weed harrowing in spring barley. They found a significant effect of row spacing only in late growth stages. The direction was found to be only significant in one out of two experiments. There was no difference for repeated harrowing between carrying it out in the same orientation and alternative orientations back and forth. They also did not find a significant effect of timing on selectivity. They suggested an effect of driving speed on selectivity but mentioned also that this needs further investigation. Lundkvist (2009) studied the effect of timing and frequency of weed harrowing on weed abundance and pre- and post-emergence weed harrowing sequences in spring cereals and peas. The combination of pre- and post-emergence harrowing provided the best control but was accompanied with yield losses of 11–14 %. For early emerging weeds pre-emergence weeding provided sufficient control, whilst for later emerging weeds pre-emergence weeding combined with one or two harrowing treatments after crop emergence was needed for effective control.

These different studies show varying results, and no general conclusions can be drawn that can be used as a guideline for whole crop-oriented weed control operations.

Duerinckx et al. (2005) investigated in a lab experiment the tine settings of a spring tine harrow to point out the effect of varied implement settings and operational conditions on the weeding performance. They looked at the mechanical actions of a tine harrow in two different soils and looked at the effects of varied implement settings and operational conditions on the tine weeding performance. The tine was pulled in soil bins without plants in order to avoid biological variances. High selectivity could be achieved with a low speed, a thin tine and a trailing or vertical tine orientation. Effective weed control however would require a high speed, a deep penetration, a standard thick tine and a leading tine operation. Since high selectivity and high efficiency need different tine settings, these settings should be based on the intended effects. Mouazen et al. (2007) mentioned that soil texture and soil physical conditions have to be taken into consideration in the adjustment of machines to realise an optimal weed control with minimal crop damage. Weis et al. (2008) described an automatically controlled real-time finger weeder developed at the University of Hohenheim in Germany. The finger weeder is to be used on experimental fields with winter cereals and summer cereals. Bi-spectral cameras make images of the crop and weeds before and after harrowing, and a soil sensor measures the soil compaction, related to resistance to mechanical action. All information is processed online and used to determine the aggressiveness of the treatment by changing the angle of the harrow tines. The adjustment is based on the highest weed control with the least crop damage. Specific details are not given.

Rasmussen et al. (2010) concluded, based on their work on the timing of post-emergence weed harrowing in Denmark and the conclusions of Rasmussen et al. (2008) and Pardo et al. (2008), that settings and use of cultivators have to be based on the immediate crop and weed response. The main reason for this is that it is impossible to predict crop and weed responses from given settings and use of implements. They mention that this is also the main reason for the development of the finger weeder of Weis et al. (2008).

2.2 Interrow Weeding

The most important innovation for interrow weeding are guidance systems that take over the steering function from the driver. With these guidance systems higher forward speeds and/or larger widths of the row between the crops can be cultivated. Pullen and Cowell (2006) mentioned that knowledge of the effect of implement geometry on the hoe path and the accuracy of a weeder steered with a mechanical steering linkage in response to a given guidance signal are very important for the development of accurate automatic steering systems for rear-mounted weeders. Most critical factors were found to be the steering ratio and the position of the hoe in relation to the follower. The performance was not significantly influenced by the longitudinal location of the instantaneous centre of rotation, the follower position, the steered wheels and the steered wheel axle.

Bonte (2011) reported on a research organisation that implemented the camera and guidance system of Garford on an existing hoe. The system performed well in grains, cabbage and sugar beets, whilst the results were varying in onions. With dry soil conditions, it was difficult for the camera system to see a difference between the soil and the onions. In grains some adjustments were necessary in the presence of wind. Merfield (2010) described the Robocrop and Eye-Drive from CLAAS AGROCOM (formerly ECO-DAN) systems as the opposite of the blind GPS steering systems. The vision-based systems can follow the crop rows and adapt to the crop growth status whilst maintaining a high capacity and weed control result. On the other hand, the blind GPS systems can control weeds close to the crop row even before the crop has emerged after seeding.

2.3 Intrarow Weeding

Intrarow weeding is a very challenging cultivation. Weeds grow also close to the plants, and the distance between the plants in the row is despite precision planting or seeding not always the same. Another complication is that it is not always easy to distinguish the weed from the crop, especially in early growth stages.

The Garford Robocrop from the UK is one of the few commercially available intrarow weeding machines. Important parts of this machine are based on the research work of Dedousis and Godwin (2005), O'Dogherty et al. (2007), Dedousis and Godwin (2008) and Tillett et al. (2008). The machine has a rotating horizontal disc with a cut-out sector. The disc moves in the row, and the cut-out sector enables the disc to pass the plants without making contact with them. The weeds are eliminated by cutting them and covering them with soil. The position of the plants in the row and relative to the rotating disc is determined with machine vision.

Dedousis and Godwin (2005) developed for this a mass flow soil dynamics model as an aid for the design of implements that control weeds by a shallow undercutting cultivation. The purpose of the model is to predict the lateral and forward displacement of soil when it is undercut by shallow working implements, as the rotating disc.

Dedousis and Godwin (2008) described the design of the rotating disc hoe. Main requirements for the design were a minimum intrarow area of 150 mm, treat weeds close to the crops with a small undisturbed zone (50 mm) and a forward speed of 1 m/s. The design of the disc was a compromise between maximum cultivated area and the tolerance to lateral and angular misalignment. Tillett et al. (2008) analysed the performance of the system in transplanted brassicas and headed lettuce with a minimum in-row plant spacing of 300 mm. The results showed a reduction of the weeds by 77, 87 and 65 % immediately after the treatment. Regrowth reduced the percentages for the first two treatments after two weeks to 74 % and 66 %, respectively. The forward speed was 1.8 km/h, which was relatively slow compared to normal intrarow cultural practice (about 4 km/h). However, more research is needed to increase the forward speed. The expected top speed is related to the maximum disc rotor speed of two plants per second (3.6 km/h for an in-row plant spacing of 0.5 m).

Dedousis and Godwin (2008) made an economic analysis of the system. They compared the disc hoe, a tractor-mounted sprayer, interrow and hand weeding and hand weeding solely. For the comparison economic cost calculator software was used; this software had over 50 implement selections. The results showed that for areas above 50 ha the disc hoe was a cheaper strategy compared to a tractor-mounted sprayer and hand weeding. For a 125 ha treated area, the calculated costs for the disc hoe were £ 81 ha^{-1}, and for the tractor-mounted sprayer and the interrow and hand weeding combination, the costs were, respectively, £ 100 ha^{-1} and £ 139 ha^{-1}.

Bonte (2011) reported the use of the Garford Robocrop in the Netherlands. A contractor uses the machine in organic maize, sugar beets and chicory. The machine performs well when the weeds are small. An advantage is also that the machine works very precise. Another advantage mentioned is that the machine does not need a GPS, which is advantageous in woody areas where receipt of GPS signals may be problematic.

Beunk (2011) reported on a contractor using a twelve-row Robocrop of Garford. The power requirement of the machine is about 60 kW (80 hp) for a four-row machine and 75 kW (100 hp) for a six-row machine. In addition the oil requirement is about 8 l/min per row. The machine of this contractor performed reasonably in sugar beets but did yet not perform well in chicory.

Nørremark et al. (2008) developed and optimised a side-shift and cycloid hoe system for intrarow hoeing. They also quantified and evaluated the performance under field conditions. The whole system relied on RTK-GPS positional information for control of the autonomous tractor, side-shift and cycloid hoe. They realised distances between tine trajectories and artificial plants in longitudinal direction between 47 ± 37 mm ($p=0.95$) and 80 ± 42 mm ($p=0.95$). The latter resulted in some non-hoed areas between uncultivated zone and tine trajectories. In transversal direction distances ranged from 0 ± 16 mm ($p=0.95$) and 17 ± 21 mm ($p=0.95$), resulting in some critical tine trajectories in the near proximity of the uncultivated zone for the latter value.

Nørremark et al. (2009) evaluated a large number of mechanical tools for the removal of weeds close to plants. Weeds that germinate close to individual crop plants have the most negative impact on crop yield. They evaluated concepts needed to have high degree of selectivity and were evaluated against a set of ten criteria. High-precision tillage and thermal weed control by laser were found to be the most promising weed control concepts to operate close to the plants.

Van Evert et al. (2009) developed a vision-based system that uses textural analysis to detect broad-leaved dock (*Rumex obtusifolius L.*) against a grass background, as step towards the automated mechanical control of this grassland weed that is hard to control in a non-chemical way. Van Evert et al. (2011) described a prototype robot system that autonomously detects broad-leaved dock and, once detected, destroys the weed by a cutting device. In a field test 93 % of the weeds were encountered, and effective weed control was achieved in 73 % of the cases. In only a few cases, a weed removal action was executed whilst there was no weed present. The estimated required time to weed 1 ha with 1,000 weeds is about 7 h. This type of weeding operation (grassland inspection once a year is sufficient and a working time from May to October) make it possible to cover a large area by one robot, reducing the operating costs per ha considerably. Annual costs of the robot were estimated to be about € 10,000. The capacity is such that this robot could service the area of about five typical dairy farms (50–100 ha each). The costs are then € 2,000 per farm per year and the farmers indicated that these costs are acceptable. This analysis shows that autonomous weed control becomes feasible when enough working hours per year can be realised. Such a system then should be able to operate in different crops, preferably crops that need weed control at different times in the year and during many hours a day, preferably day and night.

Weide et al. (2005a, b) mentioned the Pneumat weeder. This weeder uses compressed air to blow the weeds out of the row. They mentioned as advantages that it can control larger weeds than a finger weeder and it can be used in crops with a larger width of the crop row since the compressed air can travel distances larger than a finger weeder can cover. For different weeder types a better performance can be realised when the steering and depth control become more accurate and the machines can be easier and quicker adjusted. In addition Weide et al. (2008) mentioned the best weeding effect is obtained when working depth, air pressure and tractor speed are tailored to each other and adjusted according to weed growth and crop growth stages. A disadvantage is the large power requirement (60 kW for a six-row machine), which is about twice the requirement of an ordinary hoe.

Some experiences from precise hoeing machines by Applied Plant Research (PPO) in the Netherlands are reported by Beunk (2011). The French Sarl Radis hoeing machine is a simple machine that uses a light barrier to detect weeds. The machine acts in the crop row and the hoe swerves when the light beam is interrupted. Therefore, it is necessary that the crop is higher than the weeds. The large trajectory of the hoe limited the driving speed of the machine to about 3 km/h. This low forward speed is a limitation for use of this machine on a large scale in the Netherlands (Bleeker 2008). Another limitation is the use in crops with an open structure, as, for example, onion, where the light beam is not regularly interrupted (Weide et al. 2008).

Applied Plant Research (PPO) in the Netherlands improved the Sarl Radis machine in several aspects. One hoe was replaced by two hoes which increased the travel speed to about 7.5 km/h, and it also increased the area covered by the hoes. The actuators were also replaced by faster ones which also increased the area covered by the hoes (Beunk 2011).

Other techniques in the Netherlands Beunk (2011) reported on are the Robovator developed by Poulsen in Denmark and the intrarow weeder of Steketee. The Robovator is a hoeing machine that is based on plant recognition and uses cameras to detect weeds. The machine is used since short time in iceberg lettuce. The camera is mounted under a hood, and lighting is used to illuminate the crop rows.

The intrarow weeder of Steketee (Hemming et al. 2011) (Fig. 10.1) uses high-resolution cameras mounted in a hood. Strong xenon lamps are used to illuminate the crop row; this makes it possible for the machine to work in sunny conditions. Crop plants are recognised based on shape, colour and location. The area between the crop plants is weeded by two pneumatically guided hoes. The minimum plant distance in the row is 6 cm. Maximum driving speed with a 10 cm in-row plant spacing is 2 km/h and increases to 6 km/h with a 50 cm in-row plant spacing.

The machines from Steketee, Poulsen (Robovator) and Garford are commercially available in NW Europe. They are targeted at precision-seeded crops and transplanted crops. As such the machines are now spreading over Europe in crops like lettuce, sugar beet, chicory and different cabbage species.

3 Applications: Physical Weed Control

This section covers the applications that have been developed and tested for physical weed control. Physical weed control is mainly used when mechanical weeding is not sufficient (Ascard et al. 2007). Ascard et al. (2007) made an extensive review

Fig. 10.1 Steketee mechanical intrarow weed control system. Cameras detect the plants and the weeds are controlled by pneumatically guided hoes

on the use of thermal methods for weed control. These methods include the use of fire, flaming, infrared radiation, hot water, steam, electrical energy, microwave radiation, ultraviolet radiation, lasers and freezing temperatures. Only a few of these (flame weeding and to some extent infrared radiation, steam and electrocution) are used commercially, usually as an alternative to herbicides or when mechanical methods are not sufficient. Some of these methods have a high-energy requirement but, on the other hand, leave no chemical residue in the field. A cost benefit analysis to compare this method with other weed control methods, technology development to reduce the costs and improve the energy efficiency and the integration at farm level are necessary for a greater adoption. However, the availability of inexpensive herbicides and their acceptability hampers research on these subjects.

The full-field methods for thermal weed control rely on the selectivity of the crop plants to withstand a temporary increase in temperature longer than the small weed seedlings nearby the crop. The between crop row and plant-specific methods can use higher amounts of energy and thereby reach higher efficacies, though detection systems are required, or only parts of the cropped field (interrow or weed patches) can be treated.

Though recently, for steaming, Melander and Kristensen (2011) investigated the effects of soil type, moisture, structure and heat duration and concluded that 80 °C soil temperature should be sufficient to ensure satisfactory weed control. Bàrberi et al. (2009) investigated the use of additional activating compounds (KOH and CaO) during steaming. On some weed species this had a positive effect on the weed control, though the amount of activating compounds that should be added has still further to be investigated. Malkomes and Zwerger (2007) used steaming and fumigation to control weed seeds and weed seedlings and investigated during a period of 19 months the amount of germinated weeds. The amount of weed seedlings was well reduced though the methods are not yet used in practical situations.

Using lasers to control weeds was investigated by Heisel et al. (2001) and recently by Gude et al. (2010). Gude et al. (2010) showed in cooperation with the Fraunhofer Institute in Germany that it is possible to put enough energy in the growth points of weed seedlings to stop their growths. The challenge is to get the laser beam in the right position with the help of microelectronic mirroring systems. Heisel et al. (2001) used the laser as a means to cut the weeds close to the soil and compared the performance to cutting with scissors. They concluded that CO_2 lasers have the potential of being used as a cutting device for weed control.

Sartorato et al. (2006) published on the potential of microwaves for weed control. Microwave heating of plant parts could overcome the risks of fire by flame weeding or the heavy loads of water carrying by steam treatments. The microwave efficiency has to be increased to make it a competitor to other thermal weed control methods.

One of the manufacturers of physical weed control technologies, HOAF Infrared Technology (Oldenzaal, the Netherlands), combined flaming and infrared technologies in one machine for weed control and potato haulm desiccation. This combination causes that proteins congeal and cells burst open and consequently plants start to wilt. The HOAF machines work on the full field or interrow. The machines of this manufacturer are sold worldwide and used in organic farming. Another physical weed control machine that is working selectively on the plants and not full field is a Poulsen machine (Hvalso, Denmark) (Patent No PCT/DK2005/000311). This flame weeder uses camera technology to detect the crop plants and has fast switching flames that turn off in presence of crop plants and in that way control the weed seedlings in between the crop plants. The system has been tested at least in sugar beet and onions and is commercially available in Denmark, Germany and the Netherlands.

4 Applications: Chemical Weed Control

4.1 Full-Field Oriented

In broadcast application spraying, advances in technology have been achieved recently in Europe. One achievement implemented in practice is patch spraying of herbicides (Gerhards and Christensen 2003). Gerhards and Oebel (2006) described the practical experiences with such a system for site-specific weed control in arable crops. The system uses a separate mapping stage with cameras, after which a map-based application stage follows on the sprayer in a patch size of 3×12 m. On these patches three different tank mixes can be applied by a modified Kverneland Rau sprayer, based on the presence of certain weeds (Weis et al. 2008). For further adoption of the patch spraying technology on weeds in practice, real-time image processing to control the field sprayer sections is required (Miller and Lutman 2008). Furthermore, they indicate that the potential financial benefits of patch spraying of herbicides are relatively small, and that the future of these systems may be driven by environmental factors, like reduction of emissions to no crop zones like waterways and nature parts alongside crop fields. In NW Europe with relatively small fields, except some areas in Eastern Germany, the benefits for inter- and intra-row spraying of herbicides are expected to be higher. At the same time introduction of automatic section control and even individual nozzle control is sold as add on to sprayers to reduce overlapping on headlands. These techniques are also used to realise patch spraying in practice.

Another system to reduce the amount of herbicides used during weed control is the minimum lethal herbicide dose (MLHD) system. In this system the label recommended dose is reduced or split based on recommendations of a decision support system (DSS). The DSS takes into account many factors that interact like weed species, weed stage, crop stage, weather and soil conditions, spray technology, formulation and economics (Kempenaar et al. 2011). After spraying the reduced dose, the photosynthesis activity of the weeds is measured by mobile handheld photosystem I or photosystem II fluorescence measurement devices (Kempenaar and Spijker 2004). Changes in leaf photosynthesis of plants can be measured 2–3 days after herbicide treatment. When the photosynthesis level is below a threshold level, the weed will be killed by the minimum lethal herbicide dose, and no repeated spray is required to reach a good control level. Measurements on 20 plants per key weed species are required to get a good idea of the effects of the spray. To date, approximately 200 mobile photosynthesis measurement devices are used within the Netherlands to apply the MLHD system in practice (Kempenaar et al. 2011).

Online measuring of crop biomass and directly adjusting the spray volume is used within the SensiSpray system (Figs. 10.2 and 10.3) developed in the Netherlands. This so-called Canopy Density Spraying (CDS) is based on measurement of crop biomass by NDVI with a GreenSeeker sensor. Based on the NDVI measured, the spray volume is adjusted up or down in accordance to the requirements of the crop and weed density. The system is mainly used in the Netherlands for potato haulm desiccation, though other applications where crop biomass is changing are foreseen as well. Between two and ten systems have been sold of different brands and are being used on commercial farms. It could be used for weed control as well, as the NDVI of weed patches is different from the regular crop.

4.2 Plant Specific and Row Oriented

Weed plant-specific and row-oriented weed control methods are mainly used in organic farming, as chemical methods are not to be used in organic farming. On the other hand, in conventional farming the weed-specific and row-oriented approach for application of herbicides is efficient and can outperform traditional broadcast

Fig. 10.2 SensiSpray system mounted on a sprayer with air support to prevent drift

Fig. 10.3 Detail of the SensiSpray spraying system. The GreenSeeker sensor is used to determine the amount of biomass, and the spray volume is adjusted by switching on and off one to four nozzles in the nozzle holder

spray application techniques of herbicides when it concerns the amount and type of herbicides used.

In Denmark and in the Netherlands, several systems were topic of research to weed plant specific by applying herbicides. These systems used micro-sprayer application techniques. The micro-sprayer concept described by Lund et al. (2006)

in Denmark combines recognition, spraying and robot technology. A robot vehicle guided by RTK-GPS along the crop seed line carries a camera for weed seedling detection as well as a micro-sprayer. The first prototype had a 126 mm spray boom with 40 needles (hypodermic tubes), divided in eight sections, each controlled by a solenoid valve, delivering the spray fluid on exact positions on the weed seedling leaves. A newer prototype had a boom of 100 mm with 20 individually controlled tubes. With this version it was possible to spray on 5×5 mm weed cells in the field.

Experiments by Sogaard and Lund (2007) showed that the average of absolute distances of the newer prototype in trial tests was ±2.6 mm in single droplet positioning. With the application of this system in practice, the amount of glyphosate can be reduced to $4\ g \cdot ha^{-1}$ for a full control of 100 weed seedlings per m^2. This is a reduction in spray volume of two orders of magnitude, compared to broadcast spraying (Sogaard and Lund 2007). To apply micro-spraying in practice, research has focused on the spray formation and spray liquid transport to the plant as well. Lund et al. (2008) mentioned that the dynamics of a spray of an on/off spray system are quite different from a conventional spray system used in agriculture. For that reason it is important to mention the biological efficacy of such a system and relate it to conventional spray technologies.

A weed control system for removal of volunteer potato plants within crop rows was developed by Nieuwenhuizen et al. (2010b). The system (Fig. 10.4) relies on detection of the volunteer potato plants within sugar beet fields by machine vision. Two cameras, visible RGB and invisible near-infrared, were used under controlled light conditions under a cover (Nieuwenhuizen et al. 2008). Real-time image processing algorithms distinguished in an adaptive manner the crop from the weed plants at square centimetre level. Directly after the detection stage, a micro-sprayer deposited droplets containing glyphosate on the volunteer potato plants. The formulation of the spray liquid was adjusted in a way that a thicker viscous fluid prevented splashing of the systemic herbicide to the neighbouring sugar beet plants. With this system up to 77 % of weed, potato plants were killed within the sugar beet crop row (Nieuwenhuizen et al. 2010a). This was accompanied with only 1 % of unwanted death of sugar beet plants. The accuracy of the spraying system was ±14 mm in longitudinal and ±7.5 mm in transversal direction (Nieuwenhuizen et al. 2010c).

Nieuwenhuizen (2009) compared the biological efficacy of different droplet densities of the developed micro-sprayer with a flat fan nozzle in the control of volunteer potatoes. The results showed that similar efficacies as a flat fan nozzle could be realised with high droplet densities (3,022 droplets m^{-2}). Lower droplet densities (622 and 1,330 droplets m^{-2}) showed lower efficacies.

The system has not yet been introduced to commercial practice in the Netherlands, as an extension in the number of weeds that should be detected has still to be made. The possibilities for adaption in NW Europe are rather high. The use of chemicals will more and more be restricted due to national and European legislation. This will increase the demand for methods that are effective but use much less chemicals by only applying what is needed to kill the weeds. This will especially be the case for the smaller crops, i.e. the crops with a limited acreage. These crops are for crop protection companies of less interest for the development and admission of crop protection chemicals.

Fig. 10.4 Weed control system for control of volunteer potatoes within rows of sugar beets. Two cameras are mounted under the hood and the volunteer potatoes are micro-sprayed with the spray unit in the rear

5 Conclusions

Overall, full-field-based methods for weed control are the most mature or market available in NW Europe. The methods based on direct interaction with individual plants are the least mature. These methods require a very precise direction of the energy in the case of flaming or steaming, the chemical in case of a chemical-based weed-specific method or the very precise control of a hoe. This is still very difficult to achieve in a continuously varying environment in open fields. The most mature is the interrow weeding. The weeding principle is used for several decades already, and new technologies enable precise steering between the rows making the principle more effective.

For the near future, non-chemical weed control methods are preferred in NW Europe. These consist of mechanical weeding assisted by advanced machine vision-guided inter- and intrarow weeders. Mechanical weed control methods have to be improved with sensors to allow for online control of the aggressiveness and performance of weeding tools. Chicouene (2007) discussed a kind of conceptual framework that deals with the factors that influence the sensitivity of the plant to being damaged and the type of damage induced by the different implements. A proper system requires the choice of the right implement, a proper adjustment of the implement and the correct time for intervention. The types of damage inflicted on each plant should be analysed, and the weeds should be classified by different forms in which they occur and the various ways the implements inflict damage on them.

Chemical weed control methods may still be required, but the application of non-chemical and chemical weed control will have to be assisted by decision support systems to increase the efficacy and handle the variability in efficacy due to changes in weather conditions. Chikowo et al. 2009 concluded that a combination of various integrated weed management techniques allows both the long-term control of arable weeds and a significant reduction in the reliance on herbicides.

References

Ascard J, Hatcher PE, Melander B, Upadhyaya MK (2007) Thermal weed control. In: Upadhyaya MK, Blackshaw RE (eds) Non-chemical weed management: principles, concepts and technology. CABI, Wallingford, pp 155–175

Bàrberi P, Moonen AC, Peruzzi A, Fontanelli M, Raffaelli M (2009) Weed suppression by soil steaming in combination with activating compounds. Weed Res 49(1):55–66. doi:10.1111/j.1365-3180.2008.00653.x

Beunk H (2011) Schoffelen op weg naar elektronische toekomst. Landbouwmechanisatie, vol 62. Agrimedia, Wageningen

Bleeker PO (2008) Schoffel Robrocrop ruimt onkruid in de rij op. http://edepot.wur.nl/53012

Bonte E (2011) Garford Robocrop. Landbouwmechanisatie, vol 62. Agrimedia, Wageningen

Chicouene D (2007) Mechanical destruction of weeds. A review. Agron Sustain Dev 27(1):19–27. doi:10.1051/agro:2006012

Chikowo R, Faloya V, Petit S, Munier-Jolain NM (2009) Integrated Weed Management systems allow reduced reliance on herbicides and long-term weed control. Agric Ecosyst Environ 132(3–4):237–242. doi:10.1016/j.agee.2009.04.009

Christensen S, Søgaard HT, Kudsk P, Nørremark M, Lund I, Nadimi ES, Jørgensen R (2009) Site-specific weed control technologies. Weed Res 49(3):233–241. doi:10.1111/j.1365-3180.2009.00696.x

Cloutier DC, van der Weide RY, Peruzzi A, Leblanc ML (2007) Mechanical weed management. In: Upadhyaya MK, Blackshaw RE (eds) Non-chemical weed management: principles, concepts and technology. CABI, Wallingford, pp 111–135

Dedousis AP, Godwin RJ (2005) Precision mechanical weed control. European Weed Research Society, Doorwerth

Dedousis AP, Godwin RJ (2008) The rotating disc-hoe-an overview of the system for mechanical weed control. In: American Society of Agricultural and Biological Engineers annual international meeting 2008, Providence, RI, pp 451–463

Duerinckx K, Mouazen AM, Anthonis J, Ramon H (2005) Effects of spring-tine settings and operational conditions on the mechanical performance of a weed harrow tine. Biosyst Eng 91(1):21–34. doi:10.1016/j.biosystemseng.2005.02.005

European Commission (2009) Directive number 128 of 2009. Off J Eur Union L 309:16

Evert FK van, Polder G, van Der Heijden GWAM, Kempenaar C, Lotz LAP (2009) Real-time vision-based detection of Rumex obtusifolius in grassland. Weed Res 49(2):164–174. doi:10.1111/j.1365-3180.2008.00682.x

Evert FK van, Samsom J, Polder G, Vijn M, H-Jv D, Lamaker A, van der Heijden GWAM, Kempenaar C, van der Zalm T, Lotz LAP (2011) A robot to detect and control broad-leaved dock (Rumex obtusifolius L.) in grassland. J Fld Robot 28(2):264–277. doi:10.1002/rob.20377

Ferrero A, Busi R, Tesio F, Vidotto F (2007) Weed control by mechanical means in organic soybean and maize. European Weed Research Society, Doorwerth

Gerhards R, Christensen S (2003) Real-time weed detection, decision making and patch spraying in maize, sugar beet, winter wheat and winter barley. Weed Res 2003(43):385–392

Gerhards R, Oebel H (2006) Practical experiences with a system for site-specific weed control in arable crops using real-time image analysis and GPS-controlled patch spraying. Weed Res (Oxf) 46(3):185–193. doi:10.1111/j.1365-3180.2006.00504.x

Griepentrog HW, Nørremark M, Nielsen H, Blackmore BS (2005) Seed mapping of sugar beet. Precis Agric 6(2):157–165. doi:10.1007/s11119-005-1032-5

Gude J, Damerow L, Schulze-Lammers P (2010) Weeding by laser application. Paper presented at the third conference on precision crop protection, Bonn

Heisel T, Schou J, Christensen S, Andreasen C (2001) Cutting weeds with a CO2 laser. Weed Res 41(1):19–29. doi:10.1046/j.1365-3180.2001.00212.x

Hemming J, Nieuwenhuizen AT, Struik LE (2011) Image analysis system to determine crop row and plant positions for an intra-row weeding machine. In: CIGR, Funabori, Japan, 19–23 Sept 2011, p 7

Kempenaar C, Spijker JH (2004) Weed control on hard surfaces in The Netherlands. Pest Manag Sci 60(6):595–599. doi:10.1002/ps.863

Kempenaar C, Lotz LAP, Snel JFH, Smutny V, Zhang HJ (2011) Predicting herbicidal plant mortality with mobile photosynthesis meters. Weed Res 51(1):12–22. doi:10.1111/j.1365-3180.2010.00823.x

Kropff MJ, Bastiaans L, Kempenaar C, Van Der Weide RY (2008) The changing role of agriculture and tomorrow's weed research agenda. J Plant Dis Prot 21:3–8

Kurstjens DAG (2007) Precise tillage systems for enhanced non-chemical weed management. Soil Tillage Res 97(2):293–305. doi:10.1016/j.still.2006.06.011

Lund I, Sogaard HT, Graglia E (2006) Micro-spraying with one drop per weed plant. Danish Institute of Agricultural Sciences, Tjele

Lund I, Jensen PK, Olsen HJ (2008) Aspects of spraying small geometric cells. In: Alexander LS, Carpenter PI, Cooper SE et al (eds) International advances in pesticide application, Robinson College, Cambridge, UK, 9–11 January 2008, Aspects of applied biology 84. Association of Applied Biologists, Wellesbourne, pp 273–278

Lundkvist A (2009) Effects of pre- and post-emergence weed harrowing on annual weeds in peas and spring cereals. Weed Res 49(4):409–416. doi:10.1111/j.1365-3180.2009.00718.x

Malkomes HP, Zwerger P (2007) Influence of soil disinfection methods on the weed seed potential and on microbial activities in the soil (Einfluss von Umweltfaktoren auf die Samenphase annueller Unkrßuter insbesondere unter dem Gesichtspunkt der Unkrautbekämpfung). Gesunde Pflanzen 59(2):77–86. doi:10.1007/s10343-007-0150-6

Melander B, Kristensen JK (2011) Soil steaming effects on weed seedling emergence under the influence of soil type, soil moisture, soil structure and heat duration. Ann App Biol 158(2):194–203. doi:10.1111/j.1744-7348.2010.00453.x

Melander B, Rasmussen IA, Bàrberi P (2005) Integrating physical and cultural methods of weed control-examples from European research. Weed Sci 53(3):369–381. doi:10.1614/ws-04-136r

Merfield CN (2010) Precision Ag for ecological farming systems. In: LandWISE conference 2010: know your farm – with precision Ag. LandWISE Inc., Havelock North, 12–13 May 2010

Miller PCH, Lutman PJW (2008) A review of the factors influencing the technical feasibility and potential commercial uptake of the patch spraying of herbicides in arable crops. In: Alexander LS, Carpenter PI, Cooper SE et al (eds) International advances in pesticide application, Robinson College, Cambridge, UK, 9–11 January 2008, Aspects of applied biology 84. Association of Applied Biologists, Wellesbourne, pp 265–272

Mouazen AM, Duerinckx K, Ramon H, Anthonis J (2007) Soil influences on the mechanical actions of a flexible spring tine during selective weed harrowing. Biosyst Eng 96(1):7–18. doi:10.1016/j.biosystemseng.2006.09.007

Nieuwenhuizen AT (2009) Automated detection and control of volunteer potato plants. Wageningen University, Wageningen

Nieuwenhuizen AT, Van Der Steen S, Hofstee JW, Van Henten EJ (2008) Real time vision detection of weed potato plants in sugar beet fields. In: Conference: agricultural engineering – Land-Technik 2008: Landtechnik regional und International, Stuttgart-Hohenheim, pp 173–178

Nieuwenhuizen AT, Hofstee JW, van de Zande JC, Meuleman J, van Henten EJ (2010a) Classification of sugar beet and volunteer potato reflection spectra with a neural network and statistical discriminant analysis to select discriminative wavelengths. Comput Electron Agric 73(2):146–153. doi:10.1016/j.compag.2010.05.008

Nieuwenhuizen AT, Hofstee JW, van Henten EJ (2010b) Adaptive detection of volunteer potato plants in sugar beet fields. Precis Agric 11(5):433–447. doi:10.1007/s11119-009-9138-9

Nieuwenhuizen AT, Hofstee JW, van Henten EJ (2010c) Performance evaluation of an automated detection and control system for volunteer potatoes in sugar beet fields. Biosyst Eng 107(1):46–53

Nørremark M, Griepentrog HW, Nielsen J, Søgaard HT (2008) The development and assessment of the accuracy of an autonomous GPS-based system for intra-row mechanical weed control in row crops. Biosyst Eng 101(4):396–410. doi:10.1016/j.biosystemseng.2008.09.007

Nørremark M, Swain KC, Melander B (2009) Advanced non-chemical and close to plant weed control system for organic agriculture. Paper presented at the international agricultural engineering conference, Bangkok, 7–10 Dec 2009

O'Dogherty MJ, Godwin RJ, Dedousis AP, Brighton JL, Tillett ND (2007) A mathematical model of the kinematics of a rotating disc for inter- and intra-row hoeing. Biosyst Eng 96(2):169–179

Oerke EC (2006) Crop losses to pests. J Agric Sci 144(1):31–43. doi:10.1017/s0021859605005708

Pardo G, Cirujeda A, Aibar J, Cavero J, Zaragoza C (2008) Weed harrowing in winter cereal under semi-arid conditions. Span J Agric Res 6(4):661–670

Pullen DWM, Cowell PA (2006) The effect of implement geometry on the hoe path of a steered rear-mounted inter-row weeder. Biosyst Eng 94(3):373–386. doi:10.1016/j.biosystemseng.2006.03.011

Rasmussen J, Bibby BM, Schou AP (2008) Investigating the selectivity of weed harrowing with new methods. Weed Res 48(6):523–532. doi:10.1111/j.1365-3180.2008.00661.x

Rasmussen J, Mathiasen H, Bibby BM (2010) Timing of post-emergence weed harrowing. Weed Res 50(5):436–446. doi:10.1111/j.1365-3180.2010.00799.x

Sartorato I, Zanin G, Baldoin C, De Zanche C (2006) Observations on the potential of microwaves for weed control. Weed Res 46(1):1–9. doi:10.1111/j.1365-3180.2006.00484.x

Sogaard HT, Lund I (2007) Application accuracy of a machine vision-controlled robotic micro-dosing system. Biosyst Eng 96(3):315–322

Tillett ND, Hague T, Grundy AC, Dedousis AP (2008) Mechanical within-row weed control for transplanted crops using computer vision. Biosyst Eng 99(2):171–178

Wei D, Liping C, Zhijun M, Guangwei W, Ruirui Z (2010) Review of non-chemical weed management for green agriculture. Int J Agric Biol Eng 3(4):52–60. doi:10.3965/j.issn.1934-6344.2010.04.052-060

Weide RY van der, Bleeker P, Molema GJ, Kurstjens DAG (2005a) Mechanical weed control in plant rows in 2005. (Onkruidbeheersing in Nederland, nut en noodzaak) [Dutch] (Mechanische onkruidbestrijding in de gewasrij anno 2005). Gewasbescherming 36(2):76–79

Weide RY van der, Bleeker P, Molema GJ, Lotz LAP, Fogelberg F, Melander B (2005b) Innovation in mechanical weed control in crop rows. European Weed Research Society, Doorwerth

Weide RY van der, Bleeker PO, Achten VTJM, Lotz LAP, Fogelberg F, Melander B (2008) Innovation in mechanical weed control in crop rows. Weed Res (Oxf) 48(3):215–224. doi:10.1111/j.1365-3180.2008.00629.x

Weis M, Gutjahr C, Ayala VR, Gerhards R, Ritter C, Schölderle F (2008) Precision farming for weed management: techniques. Gesunde Pflanzen 60(4):171–181. doi:10.1007/s10343-008-0195-1

Chapter 11
Field Applications of Automated Weed Control: Asia

Hiroshi Okamoto, Yumiko Suzuki, and Noboru Noguchi

Abstract Weed control is very important in agriculture in Asian countries. Mechanical systems are used for many agricultural tasks such as cultivating, seeding, fertilizing, and harvesting, but weeding is not mechanized enough. There is a rising demand for nonchemical or low-chemical agriculture from the viewpoint of consumer/farmer safety and environmental preservation. At present, in Japan herbicides are used mainly for weed control. Nonchemical weeding with manual operations is also done but on a small scale because labor costs are very high. For nonchemical manual weeding, farmers are forced to work long, strenuous hours. The cost of manual labor in most Asian countries is rising along with rapid economic growth. In this chapter, the detection and removal of weeds autonomously using an advanced system developed in Japan will be discussed. The goal is a robotic weed control system that can maintain efficient and effective weed control and provide economic profits to the farmer in both conventional and organic systems.

1 Introduction

In Japanese agriculture, crop production is becoming more difficult because farming is done on a small scale with high labor requirements. Labor costs are higher in Japan than in other Asian countries. As a result, prices of Japanese products are higher than those imported from foreign countries.

H. Okamoto (✉) • N. Noguchi
Hokkaido University, Kita-9, Nishi-9, Kita-ku, Sapporo 060-8589, Japan
e-mail: hiro@bpe.agr.hokudai.ac.jp; noguchi@bpe.agr.hokudai.ac.jp

Y. Suzuki
Kitasato University, 35-1, Higashi 23-bancho, Towada 034-8628, Japan
e-mail: suzuki@vmas.kitasato-u.ac.jp

S.L. Young and F.J. Pierce (eds.), *Automation: The Future of Weed Control in Cropping Systems*, DOI 10.1007/978-94-007-7512-1_11,

Large variability exists in the economies and agricultural sectors of countries in Asia with some well developed economically and industrially and others are still in development. Agriculture in China is large scale both in field size and production, while farmers in other countries such as Japan are numerous and have small operations. Currently, Japan is one of the most-developed countries in Asia, but other countries are quickly catching up. For those countries that are smaller and heavily reliant on labor to support agricultural production, they could be facing the same situation as Japan in the near future, especially as human labor costs continue to increase.

In Japan, mechanical systems are used for many agricultural tasks, including cultivating, seeding, fertilizing, and harvesting, yet advanced mechanical weeding is lacking. Currently, the main weed control techniques are mechanical, chemical, and manual labor. Mechanical weeding with cultivators is generally applied to the crop interrows, and herbicides are used in the intra-row. A limited amount of manual weeding (hand labor) is performed on a small scale.

Globally, there is an increasing demand for nonchemical or reduced-chemical (e.g., organic) agriculture products due to safety and environmental concerns by the public. In Japan, consumers are very sensitive to the quality of and safety for consuming agricultural produce. However, this demand requires more labor working longer hours to manually weed crop fields. Therefore, the limitation in available labor and high cost also restricts the availability of organic produce across the country. While concerned, the public is not willing to pay the extremely high prices for organic produce.

One approach to address these problems is the development of a robotic weed control system with sensor units that identify crops and weeds, simultaneously. A mobile platform (e.g., robot), which consists of a camera for vision and computer for analysis, could be assembled as a first step in autonomous weed control. In the future, other features could be added, such as micro-applicators with targeting capabilities, RTK for navigation, and GIS software for mapping.

In this chapter, the detection and removal of weeds using autonomous vehicles developed in Japan will be discussed.

2 Position Control of Precision Weeder with Machine Vision

The philosophy of Precision Agriculture is making the right application at the right time at the right location, which also applies to autonomous weed control. Currently, mechanical and manual weeding requires that the operator not damage the crop while destructively removing the weed(s). For high-powered machinery, most implements are rear-mounted, thus forcing the operator to continually check for crop damage by looking backwards instead of forwards at the crop row. An implement that automatically follows the crop row would allow the operator to concentrate on steering the tractor without having to constantly look behind at the implement, which would prevent crop loss and maintain efficient weed control.

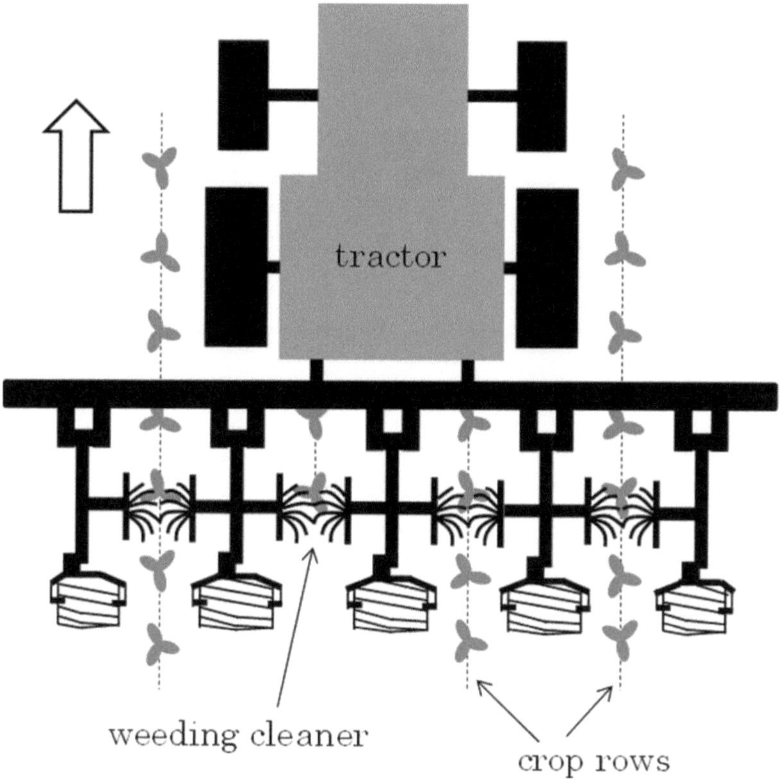

Fig. 11.1 Cultivator with weeding cleaners

A crop-row detecting system that has been developed (Okamoto et al. 2002) is composed of readily available devices such as a CCD video camera, a video capture board, and a personal computer. A tractor or a field machine that is equipped with a video camera travels along a crop row, and continuous images are captured by the video camera. The developed crop-row sensor was applied to the automatic row-following control system for the precision weeding cultivator (Nichinoki Seiko Inc. NAK-5) and tested in the farm field (Fig. 11.1).

This system was composed of the crop-row sensor and the row-following control unit (Fig. 11.2). The CCD camera of the crop-row sensor was mounted on the cultivator and took the backward crop-row images. The sensor detected the target crop-row lines and calculated the line location in real time after capturing the crop-row image. The row-following control unit received the offset value between the target row and the weeding cleaner. According to the detected offset, the hydraulic cylinder moved the whole weeder either right or left to minimize the offset.

While the system is working, the position of the weeding cleaner has to be adjusted to the target crop row essentially. However, the hydraulic cylinder moves the weeder slowly, and the system is traveling. Therefore, if the system were controlled according to the offset based on the position of the weeding cleaner, the

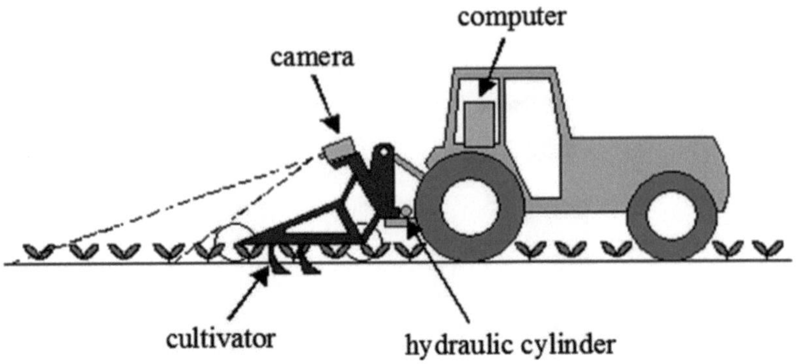

Fig. 11.2 Row-following control system

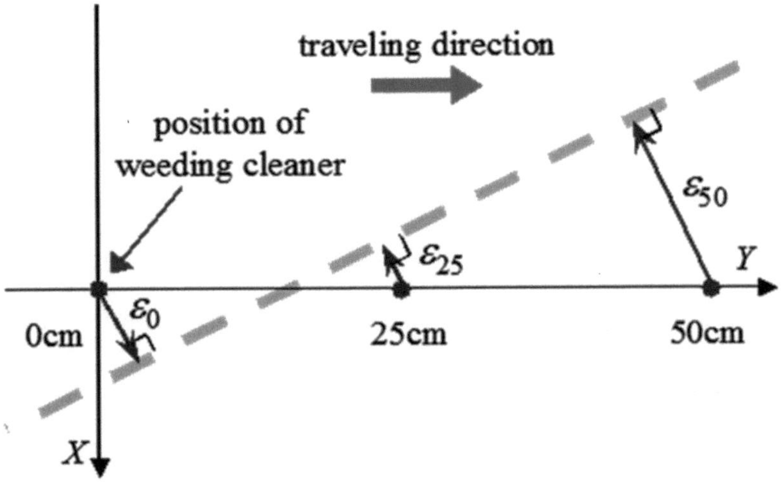

Fig. 11.3 Datum point of offset

control would be delayed. In order to avoid the delay in control, the system has to be controlled by predicting offset. Therefore, the datum point of the offset was dislocated ahead. In the field test, the datum point of the offset was set at 0, 25, and 50 cm ahead from the weeding cleaner (Fig. 11.3).

In the field test, the weeding cultivator was driven at the traveling velocity of 0.9, 1.1, and 1.3 m/s. The tested fields each had the distance of 150 m. The test was also conducted without row-following control. The cultivator was not moved along the crop row when the system was not controlled (Fig. 11.4). The cultivator was well adjusted to the target row when the system was controlled.

The system was obviously effective in precise row following (Table 11.1). In addition, by dislocating the datum point of the offset ahead, namely, predicting the offset, accuracy of adjustment to the target crop row was improved. As the traveling velocity was increased, predicting of the offset was more effective. And the accuracy of adjustment was almost equivalent when the datum point of the offset was at 50 cm.

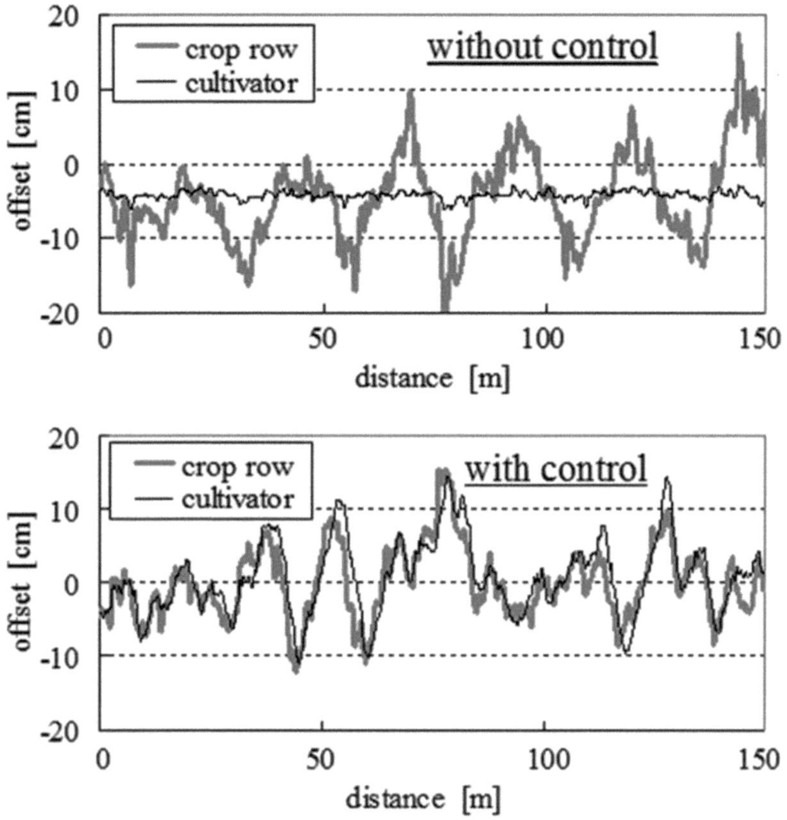

Fig. 11.4 Tracks of cultivator and crop row

Table 11.1 R.M.S. of row-following errors

	R.M.S. of row-following errors [cm]			
	Datum point of offset [cm]			
Traveling velocity [m/s]	0	25	50	Without control
0.9	2.72	2.63	2.50	5.69
1.1	3.12	2.15	2.60	8.09
1.3	4.25	3.46	2.39	6.19

3 Weed Detection with Hyperspectral Imaging

A number of studies have been carried out on image processing for weed detection. In Japan, many of the studies were conducted using an RGB color camera to detect weeds. An RGB camera can obtain spatial area information, but the images include only three low-resolution wavebands (red, green, and blue). Some of the studies used a spectrophotometer, but this instrument can obtain spectral data at one spot only and cannot obtain spatial images.

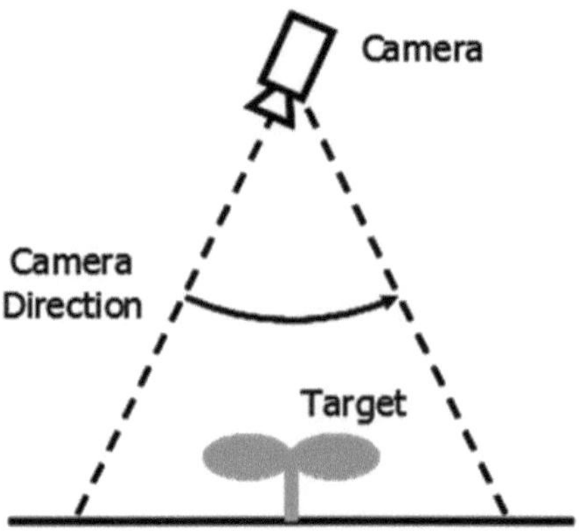

Fig. 11.5 Portable image capturing system with electric driven pan head

Recently, hyperspectral imaging was also employed for weed detection. A hyperspectral image contains both spatial and spectral information. Each pixel in a spatial hyperspectral image includes high-resolution and high-dimensional spectral data. It is considered that hyperspectral image analysis using a combination of both spatial and spectral information achieves more accurate weed detection.

A portable hyperspectral imaging system has been developed by Okamoto et al. (2006a, b, 2008) for weed detection in cropping systems. The technique is expected to be applied to future automatic mechanical weeding systems. A hyperspectral camera (ImSpector V10; Specim, Oulu, Finland) was used to acquire hyperspectral images. This camera can acquire spectral images at a wavelength range of 400–1,000 nm and a resolution of 10 nm. Two systems for image capturing were developed. The first one was the portable image capturing system, and the second one was the field-scale image capturing system. The portable image capturing system (Fig. 11.5) was used to discriminate between model crop and weeds (Okamoto et al. 2007; Suzuki et al. 2008). It consisted of the hyperspectral camera, the electric-driven pan head, and a camera tripod. In this system, the camera's optical axis was moved by an electrical motor of the pan head to scan target crops and weeds.

The field-scale image capturing system (Fig. 11.6) was used to test the weed detection methods on the field-scale crop rows and can be used for a practical weeding robot in the future. It consisted of the hyperspectral camera, a field vehicle, and a frame for camera attachment. This system scanned crops and weeds on the ground and captured images of them in large-scale field by running a vehicle. This system can acquire ground-based large-scale images with high spatial resolution. A cost of this system was low because it just needed a vehicle for running and a simple frame for camera attachment, and it did not need special and expensive implements.

Prior to discrimination between crop and weed, plant must be extracted from background soil in a hyperspectral image. The difference between plant and soil was observed in spectrum at red light range and NIR (near-infrared) light range. Normalized difference vegetation index (NDVI) was calculated from NIR, and red light range was applied to the discrimination.

Fig. 11.6 Field-scale image capturing system with running vehicle

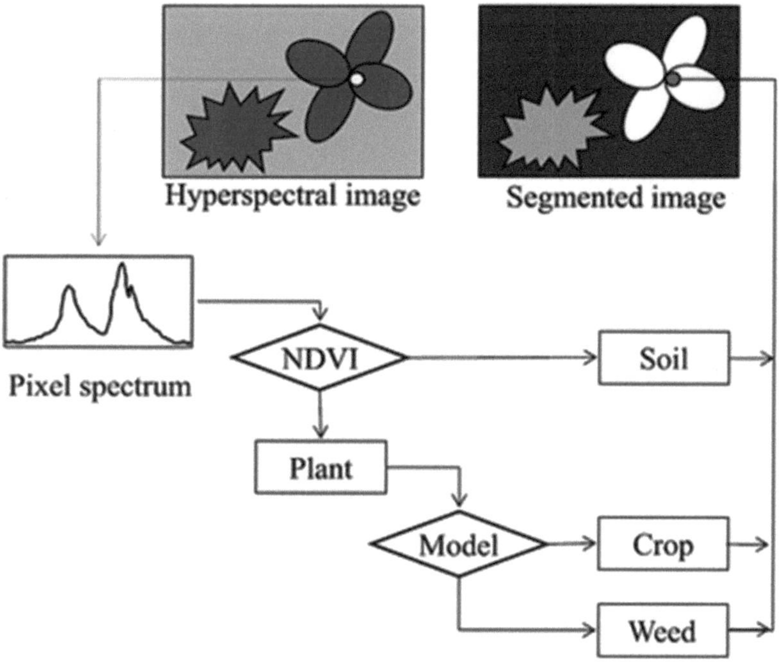

Fig. 11.7 Flow of image segmentation

It is difficult to discriminate between crop and weed by a simple threshold like discrimination between soil and crop because spectrum characteristics of crop and weed are similar. Therefore, a linear discriminant analysis (LDA) and a multilayered neural network (NN) were employed for development of a discriminator between crop and weed (Fig. 11.7). The LDA is a statistical method for discrimination between categories. The NN is a mathematical model of various parallel processing of biological neurons in the human brain. NN models with three

Table 11.2 Validation results of discrimination between crop and weed

	RAW-LDA		RAW-NN		PCA-LDA		PCA-NN	
Target	*N*	Success rate (%)	*N*	Success rate (%)	*N*	Success rate (%)	*N*	Success rate (%)
SB vs. GP	8	86.7	10	89.3	6	89.3	6	90.2
SB vs. SC	4	92.9	7	95.1	5	93.3	6	95.3
SB vs. FH	6	95.5	8	96.8	4	94.9	7	99.1
SB vs. PW	7	95.9	6	96.1	5	97.0	4	97.6

N number of explanatory variables

layers (the input, the hidden, and the output layers) were developed with a back propagation algorithm.

The field test was conducted using soybean (SB) as the target crop and goosefoot pigweed (GP), small crabgrass (SC), field horsetail (FH), and pearlwort (PW) as the target weeds. NDVI of soil samples ranged from 0.23 to 0.18, and NDVI of plant samples ranged from 0.30 to 0.77. The threshold for discriminating between soil and plant was determined to be 0.25. Success rates in the validation were 99.9 %. Discrimination between soil and plant could be performed with a high degree of accuracy.

For the top 15 wavebands selected for the models using RAW, most of the selected wavebands belonged to the green or NIR light range. It was considered that these ranges were the most important for discrimination between crop and weed. In the validation results, the success rates of the models between soybean and goosefoot pigweed were under 90 % (86.7–89.3 %), but for the other models the success rates were more than 90 % (92.9–96.8 %) (Table 11.2). The accuracy of RAW-NN models was higher than that of RAW-LDA.

The accuracy of the PCA-NN models was higher than that of the PCA-LDA model. An example of the image segmentation between soybean and each weed species can be done by applying PCA-LDA models (Fig. 11.8). In these images, pixels identified as crop (soybean) are shown in white, weed in gray, and soil in black. The majority of pixels were identified correctly. Some pixels, especially at the edges of leaves, were misidentified, and it may be difficult to discriminate completely between crop and weed. However, misidentified pixels could be removed using spatial processing such as noise reduction and majority rule; therefore, this technique will be applied to weed detection.

4 Mapping Spatial Distribution of Grasses and Weeds in Pastures

Quantity and quality of forage in pasture are not uniform because they are affected by spreading weeds, plant species, herbage mass, and plant height and density. The nutrient requirements of the grazing cattle are different by growing stage, and

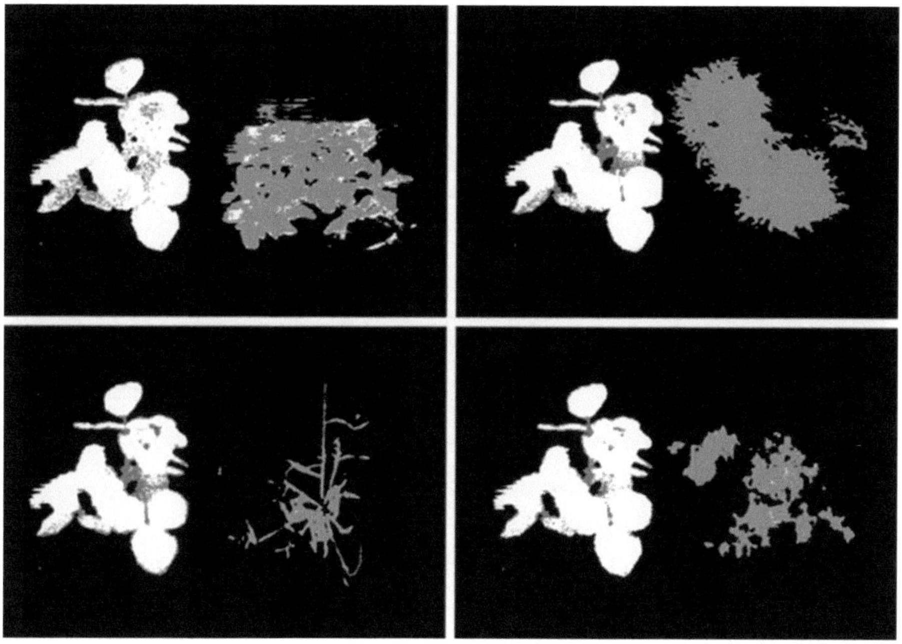

Fig. 11.8 Examples of image segmentation (crop: *white*, weed: *gray*, soil: *black*)

grazing behavior is selective. Therefore, grazing management such as grazing period, grazing time, grazing interval, and stocking rate must be correctly executed in consideration of pasture characteristics (weed spreading, quality and quantity of forage, botanical composition) to supply ample nutrients for cattle. Consequently, pasture characteristics provide essential information for achieving optimum grazing management.

Although the spatial distribution in a large-scale field was efficiently acquired using airborne and satellite remote sensing, detailed distributions could not be obtained due to low spatial resolution. On the other hand, ground-based remote sensing can acquire information with high spatial resolution, and the information for a large-scale field can be acquired by using a moving vehicle. In addition, the method can be applied for site-specific field management because it provides real-time measurements.

Ground-based hyperspectral imaging (GHI), in particular, has attracted considerable attention because detailed spectral information can be gleaned from it. In Japan, a GHI system has been developed for monitoring the spatial heterogeneity of pasture characteristics on a large-scale field (Suzuki et al. 2012). For mapping weeds in pastures, plants were first extracted from the hyperspectral image by discriminating between the plants and non-plants (bare ground and dead material) using NDVI thresholding. Next, the plants were classified into grass species and weeds. For the discrimination between grass species and weeds, the plant areas were classified into three category groups: perennial ryegrass (grass), white clover

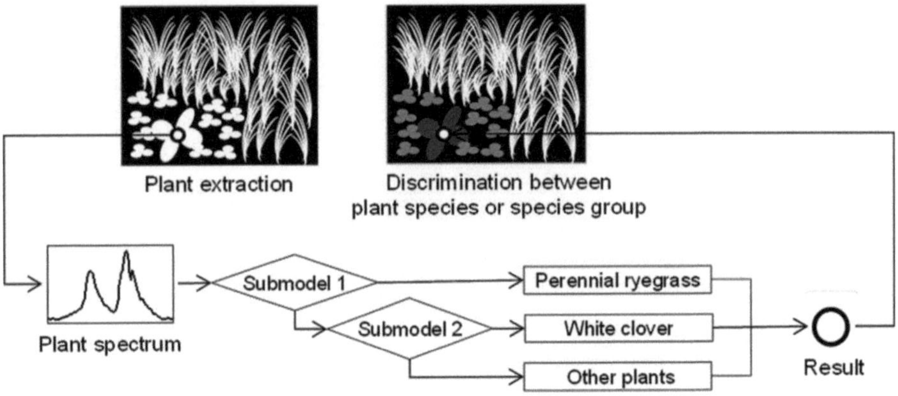

Fig. 11.9 Concept for discrimination between grass species and weeds

Table 11.3 Validation result of plant classification

Model	Selected wavelength (nm)	Success rate (%)	Overall success rates (%)
Submodel 1	741, 728, 841, 778, 753, 866	93.6	80.3
Submodel 2	753, 420, 716, 433, 580, 543, 766, 853, 568	78.2	

(grass), and other plants (weeds). In the discrimination model (Fig. 11.9), perennial ryegrass was first extracted from the plant pixels, and then the remaining pixels were classified into white clover and weeds. Therefore, two submodels (submodel 1 for extraction of perennial ryegrass, submodel 2 for extraction of white clover from weeds) were employed for graded discrimination. The submodels were developed by linear discriminant analysis (LDA) using wavebands which are selected by stepwise selection.

Most of the selected wavelengths for submodel 1, which was used for extracting perennial ryegrass, belonged to the wavelength range over 700 nm, i.e., NIR wavelengths (Table 11.3). Most of the selected wavelengths for submodel 2, which was used for discriminating between white clover and other plants, belonged to the wavelength range 420–580 nm and 716–853 nm, i.e., NIR and the visible wavelengths. The success rate of each submodel was 78.2–93.6 %, and the overall success rate was 80.3 %.

A map, which estimates the spatial distribution of grass species and weeds, was generated with the hyperspectral images which were taken by the vehicle running on the pasture (Fig. 11.10). The map reflected the actual pasture characteristics, and the spatial heterogeneity of grasses and weeds could be understood from the maps. Consequently, it was demonstrated that the ground-based hyperspectral imaging is a useful technique for estimating the pasture characteristics.

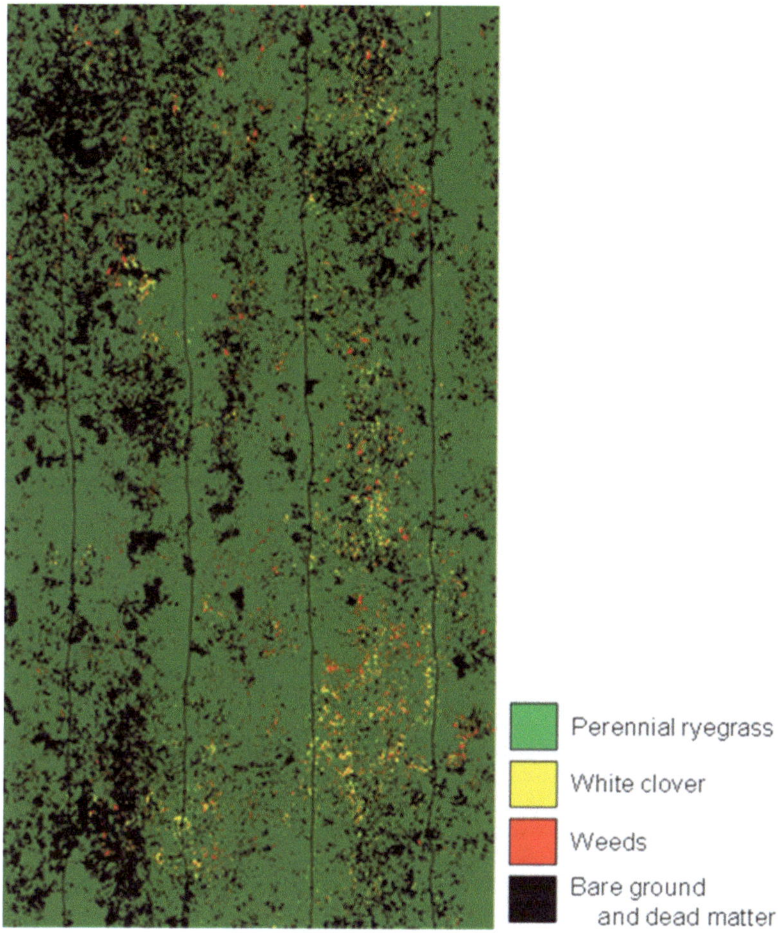

Fig. 11.10 Spatial distribution map of grass species and weeds

5 Conclusion

In this chapter, the idea of a robotic weed control system with various components was introduced. The first example was a system with position control that detects crop rows using machine vision for removal of inter- and intra-row weeds around each crop mechanically. In the second example, a hyperspectral camera specially designed for distinguishing crops and weeds was developed for removing only weeds. In the final example, a vehicle equipped with a hyperspectral imaging device was developed for determining the spatial distribution of weeds and grasses in a pasture for improved animal grazing efficiency. The information from the last example could be applied to targeted spot spraying of herbicides in an overall management plan for the pasture.

Labor costs in Japan are high, which is similar to other parts of the globe. In less-developed countries, economic growth is occurring, which is increasing labor costs. Therefore, the idea of a robotic weeding system needs to be developed to meet consumer demand and help growers be more efficient. A robotic weed control system is practical for organic, conventional, low- or high-input agriculture, since production costs can be drastically reduced as less labor is required to perform manual weed control and still maintain effective weed control. In cropping systems of the [near] future, a robot weed control system could be one of the most important components.

References

Okamoto H, Hamada K, Kataoka T, Terawaki M, Hata S (2002) Automatic guidance system with crop row sensor. In: Proceedings of automation technology for off-road equipment, Chicago, 26–27 July 2002, pp 307–316

Okamoto H, Sakai K, Murata T, Kataoka T, Hata S (2006a) Object-oriented software framework developed for agricultural hyperspectral imaging analysis. Agric Info Res 15:103–112 (in Japanese with English abstract)

Okamoto H, Sakai K, Murata T, Kataoka T, Hata S (2006b) Development of remote sensing software based on hyperspectral imaging framework. Agric Info Res 15:219–230 (in Japanese with English abstract)

Okamoto H, Murata T, Kataoka T, Hata S (2007) Plant classification for weed detection using hyperspectral imaging with wavelet analysis. Weed Biol Manag 7:31–37

Okamoto H, Suzuki Y, Sakai K, Kataoka K (2008) Unified hyperspectral imaging methodology for agricultural sensing using software framework. In: Proceeding of the international symposium on application of precision agriculture for frites and vegetables, Orlando, January 2008, pp 49–56 (2008 ISHS symposium on the application of precision agriculture for fruits and vegetables)

Suzuki Y, Okamoto H, Kataoka T (2008) Image segmentation between crop and weed using hyperspectral imaging for weed detection in soybean field. Environ Control Biol 46:163–173

Suzuki Y, Okamoto H, Takahashi M, Kataoka T, Shibata Y (2012) Mapping the spatial distribution of botanical composition and herbage mass in pastures using hyperspectral imaging. Grassl Sci 58:1–7

Section V
Economies for Automated Weed Control

Chapter 12
Economics of Technology for Precision Weed Control in Conventional and Organic Systems

Florian Diekmann and Marvin T. Batte

Abstract Regardless of the crop production system used, weeds must be controlled at or below an economic threshold in order to achieve an acceptable level of profitability. The best method of weed control will depend on a number of factors, including labor, fuel and machinery cost, crop prices, and farmer's willingness to accept production risks. In this chapter, we discuss economic factors that drive innovation in precision weed control technologies for agriculture and influence producer adoption of those technologies. We present a theoretical framework to help explain the economic incentives or disincentives to adoption of these emerging technologies. We also introduce the concept of externalities – costs or benefits realized by groups other than producers – which, if internalized to farm firms through taxes, subsidies, or restrictions, may influence producer adoption of a specific technology. We conclude with highlighting a number of important farm-level economic impacts of precision weed technology adoption.

1 Introduction

Precision weed control is a set of technologies that uses spatial and temporal information to locate, identify, and manage weeds. Precision weed control takes advantage of various innovations in automation, site-specific sensing, and application technologies to manage weeds in crop fields but essentially utilizes strategies

F. Diekmann
Food, Agricultural, and Environmental Sciences Library,
The Ohio State University, 2120 Fyffe Road, Columbus, OH 43210, USA
e-mail: diekmann.4@osu.edu

M.T. Batte (✉)
Department of Agricultural Economics, University of Kentucky,
400 C.E. Barnhart Building, Lexington, KY 40546, USA
e-mail: mbatte@columbus.rr.com

S.L. Young and F.J. Pierce (eds.), *Automation: The Future of Weed Control in Cropping Systems*, DOI 10.1007/978-94-007-7512-1_12,

similar to traditional weed management approaches, including chemical, mechanical, cultural, and biological methods (Swinton 2005). With only a limited number of commercialized hardware and software tools available to date, precision weed control is still a nascent concept compared to other precision farming technologies that have been in the marketplace for some time. A precision weed control system is comprised of three basic requirements, including a weed sensing system to detect and identify weeds, a weed management model to turn knowledge and gathered information into management decisions, and a precision weed control implement, such as a sprayer with automatic section (or individual nozzle) control for variable applications of herbicides (Christensen et al. 2009). Variable rate technology for herbicide application has received much of the attention to date, largely because of the perception that it provides the most economical and environmental benefits of all emerging precision weed control technologies (Wiles 2009).

Weed control practices have evolved considerably over the last decades with development of new herbicides, herbicide-tolerant crops, and improved application technologies. Still, weed management remains a constant challenge to agricultural productivity because of the dynamic nature of weed populations resulting from the complex interactions between production practices, soil resource characteristics, and environmental conditions (Buhler 2002). Weed populations are unevenly distributed within crop fields and tend to grow in aggregated patches varying in size, shape, and density (Dille et al. 2003; Cardina et al. 1997). Therefore, uniform herbicide applications lead to unnecessary input application on areas with weed densities below economic or weed density threshold values and have been identified as a major source of inefficiency in managing weeds (Cardina et al. 1997). Many studies have documented the potential for input and cost savings when weed control is adapted to the actual spatial and temporal distribution of weeds, and a variety of technologies have been successfully developed for precision treatment of weeds (Ritter et al. 2008; Sökefeld 2010; Schroers et al. 2010; Christensen et al. 2009; Takács-György et al. 2013). For example, precision spraying systems with automatic section control have evolved to allow selective management of input applications across the spray boom in response to detected weed populations. This technology holds substantial promise for reducing input application overlap, thus saving chemicals, fuel, and time during the application of herbicides. The potential for input savings is particularly high in situations where sprayer patterns become more complex, for example, in the case of irregular shaped fields, waterways, drainage ditches, or similar obstructions. The efficiency of this technology increases even more when combined with a precision guidance system, such as lightbar or auto-steer (Sökefeld 2010; Batte and Ehsani 2006; Shockley et al. 2012). Additional input savings are possible for applications with multiple tank sprayers, direct injection systems, and intermittent spraying of full and reduced herbicide rates that allow to more selectively control weed species in the field with different herbicides and herbicide mixtures (Gutjahr et al. 2008; Schroers et al. 2010; Wiles 2009). The full potential of precision spraying of herbicides would involve micro-targeted spraying of individual weed plants with a spray material best suited to control this weed species. To be successful, this will require advances in computer vision and plant recognition software as well as continued development of sprayer technologies.

Whether precision management of inputs is economically more viable than conventional management will vary between fields depending on the crop, the inputs, prices, the cost of new technology compared to existing technology, and the yield response variability within each field. For example, expected cost savings of variable rate technologies relative to uniform treatment will increase with greater spatial and temporal variability because optimal input application rate will vary more (English et al. 2001; Roberts et al. 2006). The economic benefits of precision weed control in particular are related to the proportion of the field that is weed infested, the degree of weed patchiness, and the spatiotemporal resolution of available sampling and spraying technologies (Barroso et al. 2004). The management of spatial and temporal variability of weed populations in general will be economically viable only when then degree of in-field variability is large enough to offset the additional costs of obtaining the information and managing the differences accordingly (Forcella 1993). In particular, development of cost-effective map-based or sensor-based technologies for detecting, mapping, and controlling weeds at the required spatial and temporal resolution remains a major challenge before the technology will become an economically viable option for farmers (Andujar et al. 2011; Wilkerson et al. 2002; Christensen et al. 2009). Developing valid decision rules that guide translation of gathered information about spatial and temporal heterogeneity of weed populations, weed-crop interactions, and cost functions into site-specific management decisions is equally critical (Gutjahr and Gerhards 2010; Gutjahr et al. 2008).

As a whole, precision weed control is a fairly recent technology, and adoption is still not widespread (Christensen et al. 2009; Takács-György et al. 2013; Jensen et al. 2012). Lowenberg-DeBoer (2003) noted that adoption rates of precision farming technologies, including precision weed control, have been much slower compared to other agricultural innovations, such as Roundup Ready© technologies in corn and soybeans that were introduced during the same time period. An extensive body of literature investigates adoption of precision farming technologies since the first components became commercially available about two decades ago and identifies key factors that influence adoption decisions (see, e.g., Khanna et al. 1999; Batte and Arnholt 2003; Adrian et al. 2005; Reichardt and Jürgens 2009; Reichardt et al. 2009). Findings suggest that adoption of precision farming technologies is influenced by a broad range of factors involving farmers' socioeconomic characteristics (Khanna 2001; Daberkow and McBride 1998; Fernandez-Cornejo et al. 2001), farmers' professional experience and education (Reichardt et al. 2009; Reichardt and Jürgens 2009; Kitchen et al. 2002; Batte and Arnholt 2003), access to information and familiarity with related information technologies (Daberkow and McBride 2003; Fountas et al. 2005), attitudes and perceptions toward precision farming technology (Adrian et al. 2005), and physical characteristics of the farm (Swinton and Lowenberg-DeBoer 1998). Other important factors influencing adoption decisions include the need for a new set of managerial abilities and additional management time required to effectively use precision farming technologies for decision making (Lowenberg-DeBoer 2003; Griffin et al. 2004). Because precision farming is intrinsically information and data intensive, the complexity of farmers' information management processes and the need for specific information management skills increase substantially, and lack thereof may restrict the effective use of these technologies (Nash et al. 2009; Reichardt and Jürgens 2009; Kitchen et al. 2002).

In the remainder of this chapter, we focus on the economic factors that drive innovation in precision weed control technologies for agriculture and influence producer adoption of those technologies. We will present a theoretical framework to help explain the economic incentives or disincentives to adoption of these emerging technologies. We will also introduce the concept of externalities – costs or benefits realized by groups other than producers – which, if internalized to farm firms through taxes, subsidies, or restrictions, may influence producer adoption of a specific technology. We will conclude with highlighting a number of important farm-level economic impacts of precision weed technology adoption.

2 Innovation and Technological Change

Technological change is a *process* that occurs over time. An early conceptual model identifies three phases of technology development and implementation: invention, innovation, and diffusion (Schumpeter 1942). *Invention* refers to the discovery or development of new knowledge. Examples relevant to precision agricultural technologies would be the discovery of methods to identify an exact position on the earth's surface through triangulation using known locations (global positioning systems or GPS) and the development of computer vision technology to allow identification of crop and weed species. *Innovation* refers to application of these inventions to produce new methods of accomplishing work related to a particular production process. There have been many areas of innovation related to the discovery of satellite-based global positioning, ranging from military applications to civilian transportation management to precision farming applications. Some of these innovations are successful and lead to new technologies that are broadly adopted, and others are dismissed as not useful. In this chapter, *technology* is referred to as methods and materials of production. Technology often is embodied in durable capital assets (machines or buildings), other inputs (hybrid seeds or agrichemicals), or even combinations of inputs (e.g., Roundup© herbicide and Roundup Ready© seed). Technology, once developed, is spread among potential users through a process of *diffusion*. Diffusion of a successful technology typically is observed to follow an S-shaped curve (Fig. 12.1) where adoption first occurs for small groups of *innovators* and *early adopters* who are innovative, embrace change, and are willing to accept risk (Rogers 2003). Research generally has shown that innovators and early adopters tend to be younger, more highly educated people, and tend to have larger farm businesses (Diederen et al. 2003). As the benefits of the technology become more broadly known, adoption accelerates with adoption by the *early majority* (Rogers 2003). The *late majority* adopt more slowly, only after advantages of the new technology are clearly demonstrated. Finally, the *laggards* are the last to adopt and typically are very averse to change. To the extent

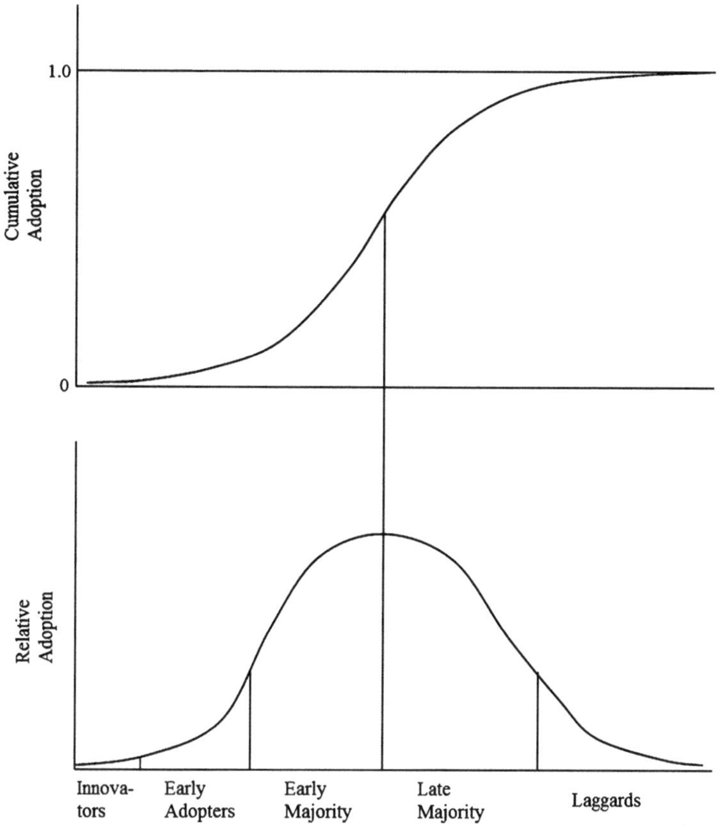

Fig. 12.1 S-shaped cumulative adoption curve (Modified from Rogers 2003)

that a technology increases outputs for a given level of inputs, profits tend to exist for those who adopt a technology early. These profits tend to diminish as larger proportions of producers adopt the technology and market prices adjust to the increased supply (Silverberg et al. 1988).

To give some perspective on how the process of invention, technology formulation, and adoption/diffusion proceed over time, let's consider the historical case of US agriculture throughout the twentieth century and explore those issues that shaped this technological change. As the century began, agriculture was characterized by draft animal power and the intense use of human labor on small farms. *Inventions* of internal combustion engines, mass production techniques, and related methods created opportunities that agricultural innovators soon recognized. The concept of a mechanical draft machine (tractor) followed, with many innovators, both farmers and industrialists, combining these new inventions into early tractors. Indeed, some of the earliest innovators were farmers who converted Model T automobiles into tractors.

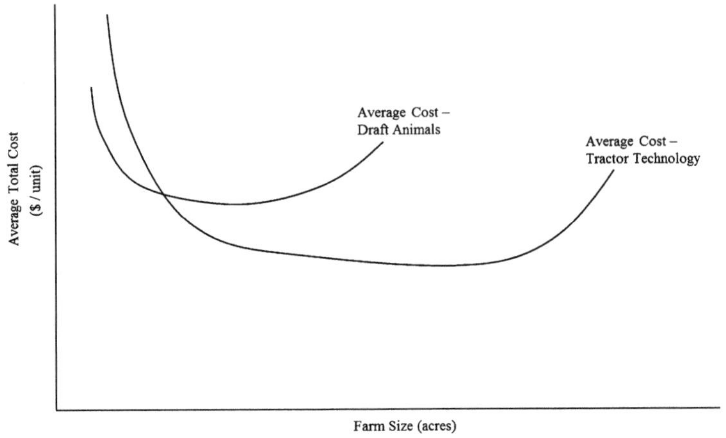

Fig. 12.2 Average total cost curves for draft animal and tractor technologies

The drive to innovate usually follows from a motive to increase the profitability of a given enterprise. Hicks (1932) first introduced the concept of *induced innovation* to help explain this drive to innovate. Following this concept within our example, farmers use a number of inputs in farming. Categorizing these into broad categories, one can think about these inputs as human labor and capital (draft animals, machinery, purchased inputs, etc.). Both input categories were required to produce a crop product using the technology of the day. However, during this time labor was increasing in cost relative to capital due to increased off-farm employment opportunities and declining machinery costs due to improvements in manufacturing. Under the prevailing farming technology, the rising cost of labor reduced the profitability of farming, and the limited availability of labor restricted farm size. Thus, farmers had an incentive to develop and adopt technology that would reduce labor needs by substituting labor-saving capital. Although the newly emerging tractor technology required substantial new capital investment, it allowed greater efficiency in the use of limited and increasingly expensive human labor. It also allowed for increased farm size. With the adoption of tractors, and expansion of the farm to fully utilize this technology, average total cost of production actually decreased (Fig. 12.2) with the adoption of mechanization relative to that of draft animal agriculture.

Induced innovation, to increasingly save relatively expensive human labor, continued over the remainder of the twentieth century, with farmers rapidly replacing existing machinery with ever larger farm machines. Figure 12.3 illustrates average cost curves for various sizes of farm equipment. Larger farm equipment allowed for increased economic efficiency (and profits) relative to smaller equipment, thereby lowering average variable costs, and, by expanding farm size to take full advantage of the larger equipment, were able to keep average fixed costs low. Thus, larger farm operators were able to produce at a lower total cost per unit of output.

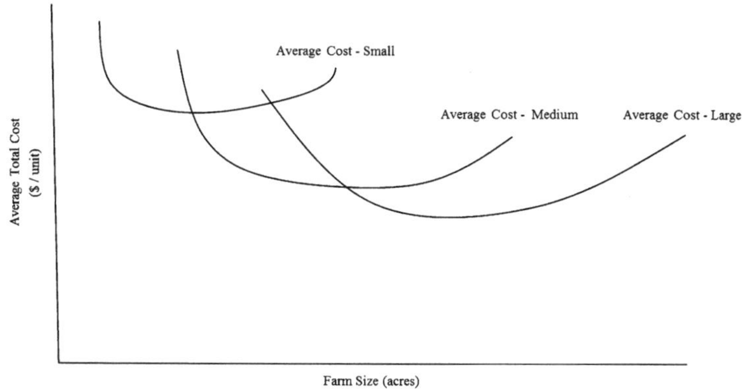

Fig. 12.3 Average total costs for farms with various sizes of machinery

Tweeten (1988) has suggested that much of the expansion of farm size that occurred during the past century was done to take advantage of larger and larger farm equipment. The introduction of other technologies such as no tillage and conservation tillage systems that reduce the number of machine passes over the field and chemical weed control technologies that eliminate tillage passes and save labor and machine operating costs have exacerbated these trends. Recently, genetic engineering has allowed crops to be modified to be tolerant of selected herbicides, further allowing substitution of chemical weed control for mechanical tillage in field crops.

3 Adoption of Emerging Weed Control Technologies in Conventional Agriculture

There is no reason to believe that motives for technology selection will be vastly different in the future than they were in the past. Induced innovation, and the desire to substitute relatively inexpensive inputs for more expensive ones, will remain as a key element driving technological change. That said, there are a number of key elements in the economic, social, and political environments that may encourage new technologies for weed control discussed in this book.

3.1 Increasing Recognition of Externalities of Production

The discussion of economics in the preceding section was based on private costs and returns to farm firms. That is, when a farmer makes a choice among competing products, or decides which technology to use, they attempt to maximize the difference in returns and costs. However, in recent decades, we have increasingly recognized that there are external costs associated with production activities. For instance, soil erosion not only costs the farmer in terms of diminished future productivity, but it costs downstream residents in terms of sediments filling streams and reservoirs, carrying fertilizer nutrients and agrichemical compounds to urban water supplies, diminishing recreational values, lowering housing values along streams and lakes, and many other impacts. These impacts are referred to as external costs. They do not impact farmers' choice of production methods unless these costs somehow are internalized. One way this can happen is if governmental agencies place a tax on farmers to offset a portion of the external costs or impose restrictions (resulting in added costs or reduced production) on producers to reduce the level of external costs. An alternative approach is to provide subsidies (external returns) tied to adoption of desired production practices. For instance, societal concerns about greenhouse gas emissions and climate change have and likely will continue to result in increased governmental regulation of agriculture. This may well translate into changes in the relative profitability of various production technologies. Should policy makers decide to encourage a particular technology, the application of an investment tax credit or other form of subsidy can dramatically alter the economics of adoption of that technology.

High technology approaches to weed control will likely fit well with an increased societal awareness (and hence policy maker concern) about adverse environmental impacts of agriculture. Spot targeted spraying techniques that apply herbicide materials only to weed plants rather than broadcast applications have significant potential to reduce off-site movement of herbicides (Wiles 2009; Burgos-Artizzu et al. 2011). Farmers may be very willing to adopt these improved weed control technologies if rewarded for this by imposing lower "externality taxes" for pollution or providing investment tax credits or other subsidies for adoption of these technologies.

3.2 Increasing Energy Costs

Since 2000, we have witnessed a significant increase in the real cost of energy. If this trend continues due to rapidly expanding demand in the developing world with slower increases in worldwide supply, then high energy prices may become an important driver of adoption of new agricultural technologies. Agrichemical inputs (fertilizers and herbicides) are highly reliant on energy. Machine vision weed control technologies allow substitution of machinery services for energy-intensive herbicide applications and may provide a profit advantage for the adopting farmer. A number of recent studies, for example, have shown the potential of micro-targeted herbicide applications to substantially reduce herbicide material application (e.g., Gutjahr and Gerhards 2010; Luck et al. 2010; Sökefeld 2010; Wiles 2009). This would be a significant cost savings for this technology, one that will increase with the rate of energy cost inflation.

3.3 *International Elements of Agricultural Demand*

Agricultural commodity markets typically are international in scope. Much of the US production of food and feed grains is sold in the work market. Yet, many of our potential trading partners have restrictions on commodity specifications that may limit our ability to sell to that market. For instance, the European Union, China, Japan, Thailand, and many other countries have placed restrictions on the use of genetically modified organisms (GMOs) in food and feed commodities. The Economic Research Service, US Department of Agriculture, estimates that 88 (93) percent of US corn (soybean) production used GMO varieties in 2012 (USDA 2013). The resistance in many foreign markets to GMOs raises concern about the long-term viability of GMO varieties in US agricultural production. Unless there is a lessening of these trade restrictions over time, farmers who produce GMO-free varieties may receive a price premium. Movement away from herbicide-resistant GMO varieties may be more feasible if micro-targeted spraying technologies and computer vision weeding technologies are available.

3.4 *Demand for Differentiated Food Products or Production Practices*

Prices of farm products are a key determinant of profitability of farm production. Prices are determined in the market as an interaction of supply and demand. However, demand is not static over time. For instance, in the case of food products, it is strongly influenced by perceptions, real or imagined, of food characteristics. Most US farmers produce commodity products: These products are not differentiated among farmers or by method of production. Their outputs are intermingled with the products of similar farmers, and the price is determined in the world market for such commodities. A recent trend has been the production of differentiated (non-commodity) food products, where the farm's product is somehow set apart as different from commodity crops.[1] Examples are certified organic foods, locally grown foods, GMO-free crops, beef produced without antibiotics or hormone supplements, pasture-finished beef or lamb, humanely produced animal products, or any other characteristic that consumers view as important. The decision to produce for one of these specialty markets will impact technology choice. For instance, foods that are certified to be organic cannot use chemical weed control or genetically modified organisms; GMO-free product producers cannot use Roundup Ready© or similar technologies. For these producers, improvements in mechanical weed control will be of increased

[1] Differentiated products each face an individualized demand curve. Hence, the prices of commodity corn and organic corn may vary substantially because a subset of consumers views these as different products with greatly differing attributes.

interest. Consumers also have indicated a willingness to pay extra for products that are certified to be environmentally friendly – e.g., to have a smaller carbon footprint or to use lower levels of agrichemical inputs (Gifford and Bernard 2008; Dannenberg 2009; Lusk et al. 2005). Computer vision mechanical weed control or micro-targeted herbicide application technologies will be of increased interest to farmers who may produce for these specialty markets.

3.5 Crop-Specific Impacts

Each crop has a unique production function and each responds differently to weed pressures. Thus, the economics of weed control differ for each crop. With all else equal, the greater the relative impact of weed competition on crop yield, and the greater the value of the crop, the higher will be the value of effective weed control. Characteristics of the crop production method also will impose restraints on the feasible weed control methods. For instance, crops grown in rows (e.g., corn, cotton) versus solid-seeded (e.g., alfalfa) or narrow row (e.g., soybeans, wheat) cropping methods create very different opportunities/constraints for tillage-based control methods. On the other hand, these characteristics may create opportunities for innovation and application of new methods. For example, machine vision combined with plant species identification may allow for mechanical weed control or micro-targeted spraying even in solid-seeded crops where tillage previously was not feasible.

3.6 Farm Size Issues

US farms range widely in size, from very small, part-time businesses to large businesses that may be international in scope. Farm size has frequently been shown to be an important determinant of technology adoption, with larger farm business operators typically being much more willing to innovate and adopt new technologies (Diederen et al. 2003; Fernandez-Cornejo et al. 2001; Sunding and Zilberman 2001). From a purely economic standpoint, larger farms have greater ability to spread the fixed costs of a new technology over greater amounts of output, resulting in lower costs per unit of production. These fixed costs arise from investment in capital assets (especially true for technologies that are embodied in machinery) or the development of special knowledge or the hiring of specially trained workers to operate the technology. Research has also shown that the operators of larger farms also tend to be more willing to innovate and to take on risks associated with technological change than operators of smaller businesses (Sunding and Zilberman 2001). Combined, this suggests that diffusion of a new agricultural technology will proceed much more rapidly among larger farms than for smaller ones.

3.7 Risk Preferences of Producers

Each individual differs in their willingness to accept (or to bear) risk. This affects willingness to take on business or financial risks. Business risks that might be important are those that differ among technologies. For instance, chemical control of weeds may be viewed as lower risk than mechanical weed control due to possible constraints on the timing of tillage due to weather events. Also, an extended period of wet weather may prevent mechanical tillage at key times, thus creating exposure to risks of yield reduction due to high weed pressure. One reason that Roundup Ready© technologies have been so rapidly adopted is that farmers have great latitude in timing weed control activities, thus greatly reducing yield risks associated with failed weed control. Financial risks are those that arise from the financial structure of the firm. Financial risks increase with the level of debt and other non-equity funding of the firm. For this reason, the size of the financial investment costs associated with a technology may be an important determinant of a farmer's willingness to adopt a technology. Lower investment systems, or those that can be performed by custom service providers, will have lower financial risks and thus be adopted more rapidly.

4 Adoption by Organic Farmers

In order to be certified to market agricultural products as organic, farmers must comply with a rigorous set of guidelines. Certification rules require that a highly structured production system be followed – one that does not allow use of most pesticide and herbicide materials, genetically modified organisms, and many other practices currently used in conventional agriculture.

Organic farmers currently use a variety of techniques to help control weed populations. A common practice in the US corn belt region is to use multiple cultivations prior to planting to allow sprouting of weed seeds which are then killed through cultivation, combined with a later than conventional planting date. Expanded crop rotations, to incorporate crops with different planting dates, can vary the habitat for weeds and allow better control. Cover crops with rapid growth can help starve weeds of light and nutrients and reduce weed pressures in subsequent crop cycles. For densely planted crops, hand weeding may be necessary to control weeds. While each of these practices is beneficial as part of an organic production system, they may come with substantial costs. Delayed planting dates may result in yield penalties due to a shorter growing season. More tillage passes require additional fuel and labor and may increase soil compaction. Specialized tillage tools will increase fixed machine investment costs. Expanded rotations mean that lower-profit crops may replace higher-profit crops in the rotation. Because weed control is such a challenge for organic producers, we believe that computer vision tillage-based weed control systems will be a huge advance for organic producers. Such technology will both reduce costs of weed control, will perhaps allow greater specialization in more profitable crops, and will reduce risks that organic farmers face due to weather-related poor weed control.

5 Farm-Level Economics of Adopting Precision Weed Control Technologies

From the previous discussion in this chapter, it becomes clear that adoption of precision weed control practices may have significant economic impacts at the farm level. Our focus in this chapter is on private costs and benefits associated with adoption of emerging weed control technologies, but it should be noted that this technology is likely to have economic implications beyond the farm level. For example, it seems feasible that spatial and temporary information and knowledge gathered at the farm level could be utilized to manage agricultural systems at local or regional scales for reduced weed interference with field crops (Swinton 2005; Maxwell and Luschei 2005). A major focus of the economic studies conducted so far in the context of precision weed control has been to investigate whether production costs per unit input can be decreased to make it more cost-effective than uniform application (Swinton 2005). Unit costs can be reduced by lowering operating input with the same output level (yield) or by increasing the output level while maintaining the same operating inputs. For example, weed control costs can be lowered by decreasing the amount of herbicides applied or by improving the placement of herbicides, thereby maximizing the net return of herbicide applications. Adoption of precision weed control technology has the potential to both increase revenues and/or lower costs at the farm level.

In the absence of subsidies or other extra-market payments to the farmer, the only source of returns to the farmer for adoption of precision weed control technologies is in the value of the crop. Total gross receipts to the cropping enterprise are the product of crop yield, price, and the number of acres harvested. The specific farm situation to which the technology is applied and how the individual farmer chooses to manage the technology will determine the impact on each of these parameters. The efficacy of weed control in most conventional cropping systems is already high, generally providing good protection against yield losses through weed interferences (Gianessi and Reigner 2007; Oerke 2006). The prospects of precision weed control to increase average yields in a significant way, therefore, are likely limited (Swinton 2005). Precision application of herbicides may mitigate negative yield effects often associated with herbicides misapplied to low weed densities areas (Weis et al. 2008; Ritter et al. 2008; Donald 1998). Oebel and Gerhards (2006) reported that site-specific herbicide applications may increase average yields in low weed density areas in cereal crops. Cedergreen (2008) found that herbicides applied at low doses can stimulate crop growth. Depending on dose-response relationships at different levels of competition between crops and weeds, precision weed control supports application of herbicides below label rates with the goal of reducing weed competitiveness sufficiently for crops to subsequently suppress weeds completely (Swinton 2005).

Precision weed control can also impact crop yields indirectly, as shown by Deike et al. (2005) who reported an increase of nitrogen efficiency under reduced herbicide applications in fields with low weed infestations. Precision management practices that enhance crop vigor (e.g., by improving nutrient and water availability for crops) may also strengthen crop competitiveness and, in turn, mitigate yield losses to weed interferences (Buhler 2002; Swinton 2005). Weed distribution patterns are often associated with soil properties, agronomic practices, and environmental variation (Buhler 2002). Because weed distributions and densities tend to vary between years and crops, input applications will need to be adjusted to the particular situation. It appears reasonable, therefore, to assume that with adoption of a technology that allows precision regulation of multiple inputs (e.g., variable rate applications for fertilizers and herbicides), average yields in some fields will increase while in others it will decrease.

Price received clearly impacts gross receipts. Crop quality for most crops is not expected to change sufficiently to impact price. However, as we discussed earlier, commodity prices may increase if crops are grown that have special characteristics, such as non-GMO, herbicide-free, or organic certification that command premium prices in the marketplace. Precision weed control technologies can help to at least partially offset higher weed management costs for these crops and will also provide the ability to preserve the identity of these crops during and after harvest. Note, however, that this is not an automatic consequence of the technology, but rather is an opportunity afforded to the adopter.

Much of the economic focus of precision weed control and other precision farming technologies to date has been on the potential for cost reductions. There are two broad categories of costs that must be borne by the farmer. Variable costs are a function of the level of output of the farm. Fixed costs represent those inputs that are invariant with the level of production. The adoption of a precision weed control system is expected to result in changes in both variable and fixed cost categories. Typical expenditures associated with precision weed control technology include output-related variable costs, such as expenditures for spatial information, data acquisition and processing, and chemical and cultural weed control, and fixed costs associated with purchase and ownership of precision weed control equipment. Examples of common fixed costs are depreciation, interest on investment, as well as taxes, insurance, and storage costs associated with yield monitors, computers and software, GPS equipment, variable rate technology application equipment, and other necessary equipment (Mooney et al. 2009). Farmers may choose to use financial leasing or custom hire in lieu of ownership for some durable assets. In the case of a financial lease, for example, the farmer will face a fixed financial lease payment instead of a charge for depreciation and interest on invested capital. A farmer may also elect to custom hire weed control services instead of owning equipment. Generally, such operating leases are offered on a variable cost basis, priced per acre or per day of operation. Depending on the implemented technology, expenditures for map-based or sensor-based (online) information and data acquisition technologies for boundary mapping, soil and site surveys, and the establishment of base and prescription maps need to be considered (Roberts et al. 2006; Rider et al. 2006). These items can incur substantial expenditures and should be considered durable investments with their costs being amortized as a fixed cost over a number of years. There are also costs associated with management time and the development of human capital that are often not considered in economic assessments for precision farming technologies (Lowenberg-DeBoer 2003). Labor and management are inputs that will not be regulated by the adopted technology, but rather are required inputs. Given the current state of the technology, we expect that human intervention in the data acquisition and decision processes will remain an important component in the foreseen future implying a substantial time commitment on part of the farm manager. Accordingly, these input costs which have been cited as an important impediment to adoption of precision farming technologies will, at least in the short term, increase with adoption of emerging weed control technologies (Lowenberg-DeBoer 2003; Fountas et al. 2005; Reichardt and Jürgens 2009). Herbicides represent a major share of crop production costs in conventional cropping system, and the monetary gains resulting from precision application based on spatial and temporal distribution of weeds, soil types, and other parameters have been well documented (see, e.g., Berge et al. 2008; Sökefeld 2010; Oebel and Gerhards 2006; Gutjahr and Gerhards 2010; Wilkerson et al. 2004; Wiles 2009; Luck et al. 2010; Dammer and Wartenberg 2007; Ritter et al. 2008; Takács-György et al. 2013). The actual amount of inputs applied under practical field conditions, however, will always be a site-specific decision. As discussed earlier, optimal input application rates depend on other crop management practices, weed infestation levels, soil properties, and the spatial and temporal variability of these factors and may increase from a conventional application level. Expenditures for gathering site-specific information and data will inevitable rise relative to the uniform application strategy regardless whether map-based or sensor-based approaches are adopted (Mooney et al. 2009). Better integration of existing technologies and advances in the development of autonomous machinery for weed detection and control are promising to help reducing the information costs of precision weed control technology in the future (Pedersen et al. 2006; Shockley et al. 2012).

Profits are the difference in total receipts and total costs. As discussed in this chapter, the change in profitability with the adoption of precision weed control technologies depends largely on the circumstances for the specific farm. On the one hand, total receipts will depend on the relative increase or decrease in average yields, change in commodity prices, and possible change in enterprise size, while increases and decreases in the variable and fixed costs categories will determine the overall costs associated with precision weed control. As pointed out, all these factors will be influenced by site-specific factors. Only as we gain more experience with precision weed control technology at the farm level will the relative contribution of each of these parameters become clearer.

6 Conclusion

Agricultural technologies are changing rapidly. New innovations arise in response to changing prices and other market and non-market signals. Market prices for labor and management continue to be high relative to the cost of machinery or agrichemical capital inputs. Farmers will continue to embrace those technologies which create the greatest relative advantage in terms of financial reward to their businesses. Even though the past decades have witnessed movement away from tillage and mechanical weed control toward chemical and biological control technologies, there may be a shift over the coming decades. Weed resistance to herbicides will reduce the effectiveness of chemical control agents. Consumers are becoming increasingly health and safety conscious, and many view the increased reliance on agrichemical control agents as risky both to the consumer and to the environment. Whether or not this perception is real, it may well translate into a demand for differentiated products that are certified to be herbicide- or pesticide-free, and which may command premium prices. If this demand materializes, it will produce an additional incentive for farmers to shift toward precision weed control. Finally, technology advancement continues. Previously, many of the precision farming technologies were high cost both in terms of investment and operation. Improvements have and will continue to decrease these costs, making these technologies much more competitive relative to traditional chemical weed control. As these changes occur, we can expect to see a continual shift toward this lower cost or higher profit allocation of resources.

References

Adrian AM, Norwood SH, Mask PL (2005) Producers' perceptions and attitudes toward precision agriculture technologies. Comput Electron Agric 48:256–271. doi:10.1016/j.compag.2005.04.004

Andujar D, Ribeiro A, Fernandez-Quintanilla C, Dorado J (2011) Accuracy and feasibility of optoelectronic sensors for weed mapping in wide row crops. Sensors 11(3):2304–2318. doi:10.3390/s110302304

Barroso J, Fernandez-Quintanilla C, Maxwell BD, Rew LJ (2004) Simulating the effects of weed spatial pattern and resolution of mapping and spraying on economics of site-specific management. Weed Res 44(6):460–468. doi:10.1111/j.1365-3180.2004.00423.x

Batte MT, Arnholt MW (2003) Precision farming adoption and use in Ohio: case studies of six leading-edge adopters. Comput Electron Agric 38(2):125–139. doi:10.1016/S0168-1699(02)00143-6

Batte MT, Ehsani MR (2006) The economics of precision guidance with auto-boom control for farmer-owned agricultural sprayers. Comput Electron Agric 53(1):28–44. doi:10.1016/j.compag.2006.03.004

Berge TW, Cederkvist HR, Aastveit AH, Fykse H (2008) Simulating the effects of mapping and spraying resolution and threshold level on accuracy of patch spraying decisions and herbicide use based on mapped weed data. Acta Agric Scand B 58(3):216–229. doi:10.1080/09064710701593087

Buhler DD (2002) Challenges and opportunities for integrated weed management. Weed Sci 50(3):273–280. doi:10.1614/0043-1745(2002)050[0273:AIAAOF]2.0.CO;2

Burgos-Artizzu XP, Ribeiro A, Guijarro M, Pajares G (2011) Real-time image processing for crop/weed discrimination in maize fields. Comput Electron Agric 75(2):337–346. doi:10.1016/j.compag.2010.12.011

Cardina J, Johnson GA, Sparrow DH (1997) The nature and consequence of weed spatial distribution. Weed Sci 45(3):364–373

Cedergreen N (2008) Herbicides can stimulate plant growth. Weed Res 48(5):429–438. doi:10.1111/j.1365-3180.2008.00646.x

Christensen S, Sogaard HT, Kudsk P, Norremark M, Lund I, Nadimi ES, Jorgensen R (2009) Site-specific weed control technologies. Weed Res 49(3):233–241. doi:10.1111/j.1365-3180.2009.00696.x

Daberkow SG, McBride WD (1998) Socioeconomic profiles of early adopters of precision agriculture technologies. J Agribus 16(2):151–168

Daberkow SG, McBride WD (2003) Farm and operator characteristics affecting the awareness and adoption of precision agriculture technologies in the US. Precis Agric 4:163–177

Dammer KH, Wartenberg G (2007) Sensor-based weed detection and application of variable herbicide rates in real time. Crop Prot 26(3):270–277. doi:10.1016/j.cropro.2005.08.018

Dannenberg A (2009) The dispersion and development of consumer preferences for genetically modified food – a meta-analysis. Ecol Econ 68(8–9):2182–2192. doi:10.1016/j.ecolecon.2009.03.008

Deike S, Pallutt B, Christen O (2005) Langzeitwirkungen von Pflanzenschutzmaßnahmen in Bezug auf die N-Effizienz. Mitt Ges Pflanzenbau 17(17):361–362

Diederen P, Van Meijl H, Wolters A, Bijak K (2003) Innovation adoption in agriculture: innovators, early adopters and laggards. Cahiers d'économie et sociologie rurales 67:30–50

Dille JA, Milner M, Groeteke JJ, Mortensen DA, Williams MM (2003) How good is your weed map? A comparison of spatial interpolators. Weed Sci 51(1):44–55. doi:10.1614/0043-1745(2002)051[0044:hgiywm]2.0.co;2

Donald DD (1998) Estimating relative crop yield loss resulting from herbicide damage using crop ground cover or rated stunting, with maize and sethoxydim as a case study. Weed Res 38(6):425–431. doi:10.1046/j.1365-3180.1998.00110.x

English BC, Mahajanashetti SB, Roberts RK (2001) Assessing spatial break-even variability in fields with two or more management zones. J Agric Appl Econ 33(3):551–565

Fernandez-Cornejo J, Daberkow S, McBride WD (2001) Decomposing the size effect on the adoption of innovations: agrobiotechnology and precision agriculture. AgBioForum 4(2):124–136

Forcella F (1993) Value of managing within-field variability. In: Robert PC, Rust RH, Larson WE (eds) Soil specific crop management: a workshop on research and development issues. American Society of Agronomy, Crop Science Society of America, and Soil Science Society of America, Minneapolis, pp 125–131

Fountas S, Blackmore S, Ess D, Hawkins S, Blumhoff G, Lowenberg-Deboer J, Sorensen CG (2005) Farmer experience with precision agriculture in Denmark and the US Eastern Corn Belt. Precis Agric 6(2):121–141. doi:10.1007/s11119-004-1030-z

Gianessi LP, Reigner NP (2007) The value of herbicides in U.S. crop production. Weed Technol 21(2):559–566. doi:10.1614/wt-06-130.1

Gifford K, Bernard JC (2008) Factor and cluster analysis of willingness to pay for organic and non-GM food. J Food Distrib Res 39(2):26–39

Griffin TW, Lowenberg-DeBoer J, Lambert DM, Peone J, Payne T, Daberkow SG (2004) Adoption, profitability, and making better use of precision farming data. Staff paper 04–06. Department of Agricultural Economics, Purdue University, West Lafayette

Gutjahr C, Gerhards R (2010) Decision rules for site-specific weed management. In: Oerke E-C, Gerhards R, Menz G, Sikora RA (eds) Precision crop protection – the challenge and use of heterogeneity. Springer, Dordrecht, pp 223–239. doi:10.1007/978-90-481-9277-9_14

Gutjahr C, Weis M, Sokefeld M, Ritter C, Mohring J, Buchse A, Piepho HR, Gerhards R (2008) Creating decision rules for site-specific weed management. J Plant Dis Prot (21):143–148

Hicks J (1932) The theory of wages. Macmillan, London

Jensen H, Jacobsen L-B, Pedersen S, Tavella E (2012) Socioeconomic impact of widespread adoption of precision farming and controlled traffic systems in Denmark. Precis Agric 13(6):661–677. doi:10.1007/s11119-012-9276-3

Khanna M (2001) Sequential adoption of site-specific technologies and its implications for nitrogen productivity: a double selectivity model. Am J Agric Econ 83(1):35–51

Khanna M, Epouhe OF, Hornbaker R (1999) Site-specific crop management: adoption patterns and incentives. Rev Agric Econ 21(2):455–472. doi:10.2307/1349891

Kitchen NR, Snyder CJ, Franzen DW, Wiebold WJ (2002) Educational needs of precision agriculture. Precis Agric 3(4):341–351

Lowenberg-DeBoer J (2003) Precision farming or convenience agriculture. In: 11th Australian agronomy conference "Solutions for a better environment". Australian Society of Agronomy, Geelong

Luck JD, Pitla SK, Shearer SA, Mueller TG, Dillon CR, Fulton JP, Higgins SF (2010) Potential for pesticide and nutrient savings via map-based automatic boom section control of spray nozzles. Comput Electron Agric 70(1):19–26. doi:10.1016/j.compag.2009.08.003

Lusk JL, Jamal M, Kurlander L, Roucan M, Taulman L (2005) A meta-analysis of genetically modified food valuation studies. J Agric Resour Econ 30(1):28–44

Maxwell BD, Luschei EC (2005) Justification for site-specific weed management based on ecology and economics. Weed Sci 53(2):221–227. doi:10.1614/WS-04-071R2

Mooney DF, Larson JA, Roberts RK, English BC (2009) Economics of the variable rate technology investment decision for agricultural sprayers. In: 2009 annual meeting southern agricultural economics association, Atlanta

Nash E, Korduan P, Bill R (2009) Applications of open geospatial web services in precision agriculture: a review. Precis Agric 10(6):546–560. doi:10.1007/s11119-009-9134-0

Oebel H, Gerhards R (2006) Camera-guided weed control – a technology for practice. J Plant Dis Prot 20:181–187

Oerke EC (2006) Crop losses to pests. J Agric Sci 144(1):31–43. doi:10.1017/S0021859605005708

Pedersen S, Fountas S, Have H, Blackmore B (2006) Agricultural robots – system analysis and economic feasibility. Precis Agric 7(4):295–308. doi:10.1007/s11119-006-9014-9

Reichardt M, Jürgens C (2009) Adoption and future perspective of precision farming in Germany: results of several surveys among different agricultural target groups. Precis Agric 10(1):73–94. doi:10.1007/s11119-008-9101-1

Reichardt M, Jürgens C, Klöble U, Hüter J, Moser K (2009) Dissemination of precision farming in Germany: acceptance, adoption, obstacles, knowledge transfer and training activities. Precis Agric 10(6):525–545. doi:10.1007/s11119-009-9112-6

Rider TW, Vogel JW, Dille JA, Dhuyvetter KC, Kastens TL (2006) An economic evaluation of site-specific herbicide application. Precis Agric 7(6):379–392. doi:10.1007/s11119-006-9012-y

Ritter C, Dicke D, Weis M, Oebel H, Piepho HP, Büchse A, Gerhards R (2008) An on-farm approach to quantify yield variation and to derive decision rules for site-specific weed management. Precis Agric 9(3):133–146. doi:10.1007/s11119-008-9061-5

Roberts RK, English BC, Larson JA (2006) The variable-rate input application decision for multiple inputs with interactions. J Agric Resour Econ 31(2):391–413

Rogers EM (2003) Diffusion of innovations, 5th edn. Free Press, New York

Schroers JO, Gerhards R, Kunisch M (2010) Economic evaluation of precision crop protection measures. In: Oerke E-C, Gerhards R, Menz G, Sikora RA (eds) Precision crop protection – the challenge and use of heterogeneity. Springer, Dordrecht, pp 417–426. doi:10.1007/978-90-481-9277-9_26

Schumpeter JA (1942) Capitalism, socialism, and democracy. Harper, New York

Shockley J, Dillon CR, Stombaugh T, Shearer S (2012) Whole farm analysis of automatic section control for agricultural machinery. Precis Agric 13(4):411–420

Silverberg G, Dosi G, Orsenigo L (1988) Innovation, diversity and diffusion: a self-organisation model. Econ J 98(393):1032–1054

Sökefeld M (2010) Variable rate technology for herbicide application. In: Oerke E-C, Gerhards R, Menz G, Sikora RA (eds) Precision crop protection – the challenge and use of heterogeneity. Springer, Dordrecht, pp 335–347. doi:10.1007/978-90-481-9277-9_21

Sunding D, Zilberman D (2001) The agricultural innovation process: research and technology adoption in a changing agricultural sector, Chapter 4. In: Bruce LG, Gordon CR (eds) Handbook of agricultural economics, vol 1, Part A. Elsevier, Amsterdam, pp 207–261. doi:10.1016/s1574-0072(01)10007-1

Swinton SM (2005) Economics of site-specific weed management. Weed Sci 53(2):259–263. doi:10.1614/WS-04-035R2

Swinton SM, Lowenberg-DeBoer J (1998) Evaluating the profitability of site-specific farming. J Prod Agric 11(4):439–446

Takács-György K, Lencsés E, Takács I (2013) Economic benefits of precision weed control and why its uptake is so slow. Stud Agric Econ 115(1):40–46. doi:10.7896/j.1222

Tweeten L (1988) Agricultural technology – the potential socio-economic impact. In: Goldberg RA (ed) Research in domestic and international agribusiness management, vol 8. JAI Press, London, pp 183–212

USDA (2013) Adoption of genetically engineered crops in the U.S. Economic Research Service, U.S. Department of Agriculture. http://www.ers.usda.gov/data-products/adoption-of-genetically-engineered-crops-in-the-us. Accessed 15 Apr 2013

Weis M, Gutjahr C, Rueda Ayala V, Gerhards R, Ritter C, Scholderle F (2008) Precision farming for weed management: techniques. Gesunde Pflanz 60(4):171–181. doi:10.1007/s10343-008-0195-1

Wiles L (2009) Beyond patch spraying: site-specific weed management with several herbicides. Precis Agric 10(3):277–290. doi:10.1007/s11119-008-9097-6

Wilkerson GG, Wiles LJ, Bennett AC (2002) Weed management decision models: pitfalls, perceptions, and possibilities of the economic threshold approach. Weed Sci 50(4):411–424. doi:10.1614/0043-1745(2002)050[0411:WMDMPP]2.0.CO;2

Wilkerson GG, Price AJ, Bennett AC, Krueger DW, Roberson GT, Robinson BL (2004) Evaluating the potential for site-specific herbicide application in soybean. Weed Technol 18(4):1101–1110. doi:10.1614/WT-03-258R

Chapter 13
Future Adoption of Automation in Weed Control

Josse De Baerdemaeker

Abstract The future adoption of automated weed control, either chemical or mechanical or otherwise, depends on a number of driving forces as well as on constraints that affect the diffusion of innovations. Some driving forces are the high labor requirements for weed control in organic agriculture, the development of herbicide resistance of weeds, and societal pressure for reduction of chemical use in agriculture and for traceability of cultivation practices. In addition, financial stimulation by government programs may accelerate acceptance. The constraints can be either technical in terms of working speed or reliability of technology, but also the relative age of existing chemical sprayers may slow down the diffusion. A number of examples of new technology introduction in agriculture are discussed to find similarities that can be a base for forecasting the adoption rate. We find some of the constraints and drivers for technology adoption and also if and how these drivers and constraints were really effective or were surpassed by other social or behavioral phenomena. It is expected that during the initial phases of adoption of automated weed control, a number of technology advances will be made that can enhance the acceptability by farmers as well as the willingness to invest by manufacturers. A 20-year period for a substantial market share of automated weed control equipment is expected.

J. De Baerdemaeker (✉)
Division of Mechatronics, Biostatistics and Sensors (MeBioS), KU Leuven, KasteelparkArenberg 30, Box 2456, BE3001 Heverlee, Belgium
e-mail: josse.debaerdemaeker@biw.kuleuven.be

S.L. Young and F.J. Pierce (eds.), *Automation: The Future of Weed Control in Cropping Systems*, DOI 10.1007/978-94-007-7512-1_13,

1 Introduction

Automation of weed control can involve two different paths. One is automation in chemical weed control, the other automation in mechanical weed control. The driving forces behind both may to a certain extend be similar. There is a societal pressure to reduce the chemical use in agriculture. This can be done by more accurate application of chemicals only where required and at a correct and effective dose. Rather than whole-field application, it involves detection and recognition of weeds followed by application of the most effective chemical component at the required dose. The application may use prescription maps based on weed locations from a detection and recognition step that can involve land based as well as aerial detection systems. Or detection and recognition may be installed on the same equipment as used for application. Automation of chemical weed control faces the risk of weeds developing resistance. New chemical molecules to overcome the resistance to current herbicides are not readily available, and it is likely that in the future for controlling more and more weed, species-specific chemicals will be needed, adding to the challenges for automatic application.

Automation of mechanical weed control requires both detection and mechanical actuation to remove the weed. There is also the risk of mechanical damage to the crop. The dynamics of the actuators and cutting tools largely affect the working speed. The weight of the equipment and the forces involved may pose severe limitations on the working width. Both methods, chemical and mechanical, rely on detection and recognition, and the success of these may depend on the growth stage of both weed and crop. Furthermore, the success of the weed control action itself is also likely to depend on the growth stage as well as on the weather conditions during and after applications.

Harker and O'Donovan (2013) defined integrated weed management (IWM) as the use of more than one weed management tactic (biological, chemical, cultural, or physical) during or surrounding a crop life cycle in a given field. They also stated the following: "*One might inappropriately conclude that IWM implementation means that herbicides should be avoided in preference for other weed management methods. However, IWM should not be about the exclusion of one method for another as much as it is about overall technique diversity. Any weed management method that is continuously repeated provides heavy selection pressure for weed adaptation and resistance to that practice. Intense and continuous barnyard grass [Echinochloa crus-galli (L.)Beauv.]hand-weeding in rice (Oryza sativa L.) allowed the selection of rice-mimic biotypes that "resisted" hand-weeding efforts* (Barrett 1983). *Therefore, weeds will likely resist any often-repeated weed management technique. In an IWM program, using a diversity of weed management methods is more important than striving to exclude any single method.*" Currently, most cropping systems, organic and conventional, lack true integration of weed management techniques due to spatial and temporal constraints. There is an excessive reliance on a single tool (e.g., herbicides or a chosen physical weed control) that puts selection pressure on the weed population conferring resistant biotypes, which eventually spread through the entire area rendering the tool useless. An alternative is automated weed management, where a decision support system moves through the crop field on a platform that simultaneously identifies weeds and activates an appropriate tool (e.g., micro-spray, micro-disc, micro-flame, and micro-laser). In this way, the decision is to use a selection of "little hammers" all at once, thereby allowing for weed management with no risk of resistance development to any single technique.

The above-mentioned considerations can affect the acceptance and diffusion of the new technological developments depending on the importance placed on by farmers, manufacturers, consumers, and governments. The rate of adoption of new technologies is in many cases studied on the basis of dynamic models. The models are then calibrated on historical data. However, the calibrated models are not readily transferred from one technology or activity to a new one. Nevertheless, they are briefly treated in the following paragraphs. Thereafter, it will be attempted to look for similarities in technology that can help to evaluate the adoption rate of automated weed control.

2 Technology Adoption Models

Widespread adoption of new technology is not instantaneous. Once initial adoption has started, the diffusion is more or less S-shaped based on four questions about general technology adoption forecast: How sure? How much? How soon? and How fast? (Vanston 2008) (Fig. 13.1).

This S-shape may be delayed between regions or countries, and the shape may also be different, indicating a slower or a more rapid adoption. In the time course or different phases of the adoption rate (Fig. 13.2), innovators, early adopters,

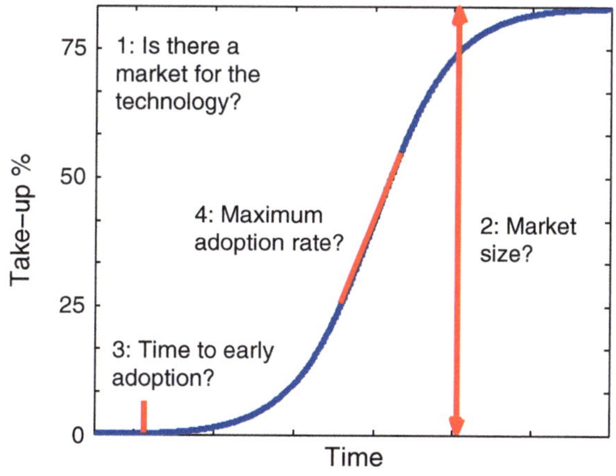

Fig. 13.1 Questions for general technology take-up forecast

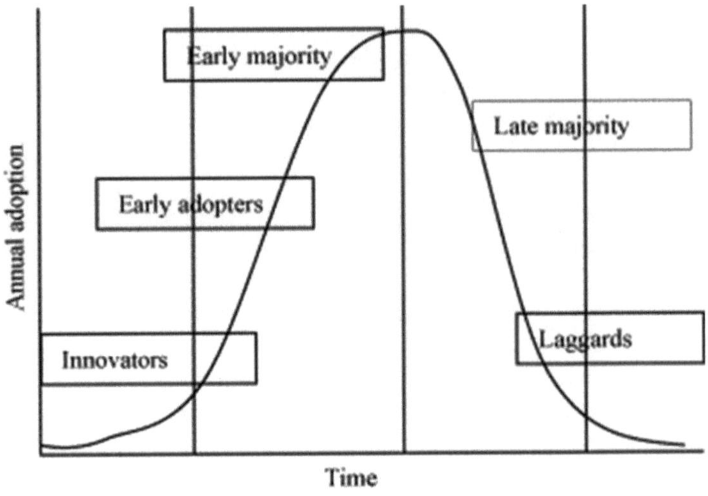

Fig. 13.2 Diffusion curve: adopters vs. time (UshaRao and Kishore 2010)

early majority, late majority, and laggards can be distinguished. UshaRao and Kishore (2010) give a review of technology diffusion models with reference to renewable energy technologies.

As stated earlier, these models are usually calibrated on historical data of mother technology adoptions. However, looking at the future adoption of new technology, they do not provide sufficient reliability as conditions for each technology may be different. Questions arise, such as will it likely be driven by replacement factors (i.e., existing units breaking or wearing out), substitution (i.e., new technology is superior causing premature change-outs), or diffusion (i.e., early adopters teach late adopters). This leads to questions about the drivers for adoption of new technology, such as how do their strengths compare to the constraints, can constraints be overcome, and is there a balance between drivers and constraints that is likely to change in the future? Since new technology addresses some expected social, economic, or ecological evolutions, competing solutions must also be considered. Is standardization required or is other technology availability a prerequisite for introducing the new technology: Is a completely new system involved or is it incremental technology? Those who must face these questions include industry as well as farmers. This also leads to questions about who will financially benefit from the new technology or who is likely to face financial losses. Even when prospects for adoption by end users may be good, manufacturers also have to consider the investments and risks involved in developing and engineering the new technology and setting up the production lines, which will be discussed in this chapter. Vanston (2008) gives some practical tips that can be used for forecasting new technology adoption.

3 Examples of Adoption of New Technology in Agriculture

Steele (2009) looked at data on the diffusion of two recent agricultural innovations in the twentieth-century United States: the adoption of hybrid corn and the tractor adoption. As for hybrid corn, he refers to a study by Griliches (1960) who has argued that the observed delay reflects initial preferential targeting of the most profitable regions by seed corn producers, and an additional factor relating to regional variation in the allocation of effort on selective breeding of hybrids adapted to local conditions by agricultural experimental stations in different states. In addition, the adoption rate of hybrid corn also reflected the expectations of yield gains, which were higher in areas where absolute yield was already higher. He concluded that temporal and spatiotemporal patterns of innovation diffusion do not, in themselves, enable the diagnosis of the importance of different kinds of bias in the decisions of adopters about the optimal timing of adoption. According to Griliches (1960), his study had at least one interesting implication. "*Hybrid corn was an innovation which was more profitable in the "good" areas than in the "poor" areas. This, probably, is also a characteristic of many other innovations. Obviously, tractors contribute more on large than on small farms, and so forth. Hence, there may a tendency for technological change to accentuate regional disparities in levels of income and rates of growth.*"

3.1 Tractors

Looking at the diffusion of the tractor adoption across the United States, Duffy-Martini and Siberberg (2004) comment that the utility of adoption of the tractor was not the same for all farming systems. For this reason and also because of the data availability, they restricted the study to farms in Iowa from 1920 to 1940. "*The diffusion of the tractor was not uniform across the US. As the tractor developed, various farming systems found them useful albeit in different decades. Very roughly, the small grain region and Far West were the earliest adopters. This farming system has an annual cycle of plow, plant and reap. Although the planting was usually done with horses, plowing and reaping could be done with tractors. The huge power needs of these tasks made the tractor a valuable addition to the farm and reduced the number of horses needed. The more complicated farming system of the Corn Belt has a cycle of plow, plant, cultivate and hand harvest for crops. Because of smaller acreages and cultivation between the rows of corn, tractors did not displace many horses until the introduction of the Farmall tractor in 1925. The Cotton Belt did not find tractors competitive until the mechanical cotton picker allowed for the displacement of labor and horses in harvest. The eastern states adoption was more individual as the evolution of the tractor and specific implements became available. Thus it is important to segregate the type of farming by region or even state.*" Nevertheless, they argue "*that the main reason why farmers switched to tractors was because tractors provided farmers with more time to raise livestock or to work off the farm, not because tractors lowered the cost of raising crops. Our results demonstrate that the seemingly slow rate of tractor adoption was in fact wealth-maximizing, and due largely to the improvement in implements which occurred late in this period.*"

3.2 Milking Robot

Robotic milking research made a proof of concept in the 1980s. In 1992, Lely introduced the Lely Astronaut milking robot in the market. At this moment there are several global producers of milking robots. Snapshot Survey 2011 by the European Dairy Farmers (EDF) and AgriBenchmark asked future-oriented milk producers in 20 countries about current and planned changes in milking strategies. Back in 2006 only 4 % of survey participant farms were using the milking robot. By 2016 the respective percentage is expected to be 22 % (NN 2012). According to Francisco Rodriguez (2012), "*it is expected that at around 2020 50 % of the cows in the UK will be milked by a robot. The robot may be an expensive durable investment, but it relieves the farmers and farm workers of the burden of 7 days presence in the milking parlor. A growing lack of high-quality, affordable labor, in combination with the demand for higher efficiency, lower costs and flexible lifestyles, has created a need for robotic milking and other automated systems on dairies. Without a motivated, competent and committed workforce, it's impossible to build a successful dairy business in today's volatile market environment. At the same time, dairy families, owners and employees – especially new generations – are actively more social and mobile, seeking a new kind of flexibility in their work, a profitable business and a better quality of life.*" This is similar to conclusions about tractor adoption of Duffy-Martini and Siberberg (2004) stated earlier.

3.3 Renewable Energy Technology

In a review of technology diffusion models with reference to renewable energy technologies, UshaRao and Kishore (2010) also give some comments on actual adoption versus the potential for adoption, which are summarized hereafter. The demand for renewable energy technologies (RETs) is being created by several governments through a set of policies, incentives, and regulations. These incentives are deemed necessary due to their inherent characteristics of RETs such as high upfront costs, lack of level playing field but distinct advantages from energy security, and environmental and social considerations. This is illustrated by statements for the United Kingdom market such as "*There has never been a better time to consider significant investment opportunities that are now available to help ease the financial burden that have previously prevented the increase take up of renewable technologies*" (NN 2013). Lewis and Wiser (2007) examined the importance of national and subnational policies in supporting the development of successful global wind turbine manufacturing companies. They focused on 12 countries: Denmark, Germany, Spain, the United States, the Netherlands, the United Kingdom, Australia, Canada, Japan, India, Brazil, and China. All of these countries have either fostered or are attempting to foster the development of a domestic wind technology manufacturing industry, though to varying degrees. The authors show that virtually all of the leading wind turbine manufacturers come from countries that have historically maintained strong policy environments for wind development.

It follows that RET applications mainly originated in the market with government subsidy and continue to attract several incentives for varying periods ranging from 5 to 15 years or more. Even after three decades of their promotion, only 20–25 % of their potential has been realized (UshaRao and Kishore 2010). Technological changes and improvements and budgetary pressure may result in government policy changes over periods much shorter than the expected technological life span of the initial installations. Higher uncertainty or inconsistent government policy will affect the introduction and adoption rate of technology.

3.4 Adoption of Rollover Protective Structures (ROPS) on Tractors

Government incentive and legislation can accelerate the adoption of new technology. One such case is the use of rollover safety structures on tractors. It is standard for newer tractors (presumably built after 1985) to come equipped with these structures, and it could be expected that for safety reasons (a matter of life or death) everyone owning a tractor would also install or retrofit one. Yet, according to Myers (2009), tractor overturns are still the leading cause of occupational agricultural deaths in the United States. They report that "*between 1992 and 2005, 1,412 workers on farms died from tractor overturns. The Roll-Over Protective Structure (ROPS) was developed to protect tractor operators from death and disability from these events by providing a protective zone for the operator during a tractor overturn. The data, which were collected for NIOSH by the U.S. Department of Agriculture (USDA), National Agricultural Statistics Service (NASS), show ROPS use has increased from 38 % in 1993 to 51 % in 2004 (an average increase of 1.2 % per year). More recent NIOSH/NASS ROPS surveillance data indicate the prevalence of ROPS-equipped tractors reached 59 % in 2006* (NASS 2008). *While the increase in ROPS-equipped tractors is encouraging, the slow rate of ROPS adoption continues to frustrate the agricultural safety community. It may seem obvious that farmers with limited resources do not have the capital to buy new tractors or retrofit their existing tractors and cost has been identified by farmers as one barrier to retrofitting ROPS on older tractors, even if they will continue to be used for decades. However, economics are not the only factors influencing reluctance to place ROPS on older tractors. Studies have shown that even with an economic incentive (like 100 % subsidy),*

ROPS acceptance by the farm operator was not 100 %." Perhaps this is more a matter of like or dislike of the protective structure rather than economic or rational motivation.

4 The Current Market of Sprayer Technology for Weed Control

In the previous section, several innovations were discussed that may have some similarities with automated weed control. The tractor or the hybrid corn adoptions were really about new technology introduction and economic wealth increase. However, the tractor had to replace horses and suitable tractor-driven implements needed to become available. Hybrid corn could just replace the conventional corn without needing new tools, just adaption of farming management. Renewable energy introductions as well as the installation of rollover protection structures are driven by considerations of environmental awareness and human health. Interestingly, this awareness is not sufficient in itself and that largely through government incentives are these technologies able to penetrate the market. The adoption of the milking robot has a large effect on the social life as well as on the management of the farm. However, this technology must replace existing milking systems, and this will be faster in case where existing installations are approaching their economic life span or becoming obsolete. In the United Kingdom in 2002, approximately 40 % of the arable land was treated by sprayer equipment less than 5 years old and the remaining 40 % between 5 and 10 years, 18 % between 10 and 20 years, and 2 % over 20 years old (Garthwaite 2002).

Similar drivers and constraints in the adoption of the above technologies exist for automated weed management adoption. They help to raise the questions (and possibly answers) when considering the introduction rate of these new technologies. The annual sales of new sprayers in the European Union are around 9,000 units in addition to 1,000 self-propelled sprayers. It can be expected that the latter are mostly equipped with the latest technologies such as GPS, automatic boom control, section control, flow rate and pressure control, or pulsed nozzles and even direct injection of the active compounds in a mixing chamber or at the nozzle. These features are incremental innovations that allow higher working speed with improved accuracy and alleviating the burden on the driver. GPS and record keeping of spray location and dose also fit in the traceability that is being demanded by government retailers and consumers. Good prices for crops in number of years make that farmers have sufficient funds to buy their own equipment. Return on investment of these features is good. These features help the farmer to be in charge of his pest control himself. He can do the treatment at the optimal moment himself rather than having to rely on custom operators.

Sprayer testing in Europe has also made farmers aware of the need for well maintained and properly operating equipment. Together with environmental regulations, it motivates the purchase of new equipment. Looking at the market evolution of conventional chemical applicators, then 25 years ago, there were many different local manufactures with local or regional component suppliers. Environmental pressure led to testing of spraying equipment, and this in turn set higher standards and requirements for sprayers that could not easily be met by all local suppliers. Nozzle testing and drift reduction requirements made that a few global suppliers of these components emerged and to a merger of smaller manufacturers. The more sophisticated equipment to reduce drift and other techniques to satisfy environmental requirements could only be met by suppliers with sufficient turnover.

Global or multinational equipment manufacturers did spot a growing market. They also are familiar with making advanced equipment and dealing with government regulations. They did in-house developments or started the production with the take-over of smaller (regional) makers. For example in the United States, the agricultural equipment engines will have to adhere to the new federally mandated emission regulations. This has perhaps an effect on the market share of companies. For several years, the gap in fleet makeup between the largest self-propelled sprayer manufacturers such as AGCO Corp., John Deere, and Case IH and the other makers has grown progressively wider (Sfiligoj 2012).

5 Current Market Situation of Automatic Weed Detection and Treatment

For automatic mechanical weed treatment or precision chemical treatment, only a few systems are currently in the market, which makes it difficult to reliably estimate the parameters of a forecasting model. Currently, the technology is developed at research institutions and produced by either spin-off companies of these institutes or picked up by small companies in specific markets. For this equipment, small companies have a business plan or model that is most likely completely different from how a multinational or global company approaches the market.

Automatic weed treatment tools remain for the moment niche products, and global manufacturers probably watch how to step in. The equipment needs easy adaptability for specific markets for which local product support is required. This implies an understanding of local agricultural production and crop and weed science. The return on investment for automatic mechanical weeding is uncertain but is probably good when used as an alternative to manual weed control.

An important driving factor in conventional agriculture may be the loss of herbicides (removal from market) in certain areas (e.g., Europe) that will increase awareness by farmers or force them to use alternative technologies compared to conventional chemical treatment. Another consideration is whether these new technologies will replace conventional sprayers or will be additional equipment on the farm. Advanced automated equipment for chemical weed control can perhaps also be designed to serve for pest and disease treatment. This is not the case for mechanical weed control.

Table 13.1 Comparison of conventional chemical, automated precision chemical, and mechanical weed control

Similarities	Differences
General crop and weed growth knowledge	Smaller working width for mechanical
Optimal timing for control	Higher cost for mechanical
Extensive testing required	Few equipment manufacturers for conventional
Specialized and local knowledge	Chemical companies contribute to HR weeds
Machine safety requirements	Automation requires less government approval
	Spacing and guidance critical for automation
	No chemicals in organic production
	Closed canopy limits mechanical
	Soil condition important for mechanical
	Weed and pest control allows multiuse for chemical equipment

6 Automation Technology Versus Conventional Chemical Weed Control

Conventional chemical, automatic precision chemical, or mechanical weed control can be complementary technologies as well as competing technologies. It is therefore interesting to look at similarities as well as differences between the technologies and their applicability (Table 13.1). This is informative for assessing the driving forces as well as the constraints for new technology introduction.

7 Automated Weed Control Technology Adoption Considerations

New technologies often have a problem developing a start to replace existing technologies because the existing technologies are (almost) optimized for current practices and have economies of scale. The new technologies require a change in cultural practices and must often start by looking for a niche in which it can grow because of a unique need. Organic production may be the spur for the growth of mechanical weed control. Since chemicals cannot be used with organic production, there is a ready market. That market may allow technologies to be evaluated and optimized, as well as manufacturers and users to develop proper methods of design, manufacturing, marketing, and use. However, in some regions, high pressure may exist to change to different cultivation methods (like no or reduced tillage farming). This in itself involves considerable change in farming practices such that opportunities or necessities may arise for novel automated weed control technology.

The adoption of new technologies is affected by perception of the potential users and economics in terms of costs both for the user and the supplier and the financial or regulatory stimuli from governments or other groups (e.g., nongovernment organizations, retailers, or consumers) (Table 13.2).

Table 13.2 Automated weed control perceptions, intentions, and economics of early adoption

Farmers	Companies	Markets
Not all weeds killed – aesthetics	Innovation in small firms	Cost for early investment
New technology – changes	Small taken over by big	New automated control market
Special qualifications/training	Large put risk to farmer	Return on investment period
Durable/quality equipment	Lower cost at large scale	Rapid change – obsolete
Trust in local dealer	Competition risk for novel	Taking economic/financial risk
Loyalty to certain brand	Test market	
Capacity to cover large area	Reliable components in poor conditions	
Ownership vs. custom operators	Adaption to or expansion of CAN bus	
Herbicide-resistant weeds	First in the market role	
Timeliness of operations	Interoperability of products	
Crop damage	Failure affordable	
Backup for failure/failed system	Farmer willing to pay	

7.1 Robustness and Redundancy of New Technology

The challenge is determining how to build robustness and redundancy into the system. The conventional chemical system has a somewhat redundant approach by competing companies having competing chemistries. In this system, redundancy and robustness are at no cost to the farmer. As weeds become resistant to more herbicides and new chemistries are not introduced, these two factors disappear.

Alternative mechanical solutions will likely involve additional cost. There may be different end effectors or actuators which cost the farmer something, even if the same sensors and structures may be used. Of course, the possibility of mechanical and chemical systems gives a high redundancy, but only when the alternative can be used after the initially selected fails. (For example, a pre-emerge herbicide is no good if a mechanical cultivation fails.)

From the viewpoint of the manufacturer, the question is who will first develop the technologies? There certainly is a lot of basic and applied research going on in institutes and universities (see other chapters in this book). Companies are doing technology scans as well as in-house research and looking for market opportunities.

There is always a learning requirement for the introduction of new technology. In case this new technology is desirable for the environmental effects, to have healthier products in the market or to stimulate new production methods, then incentives besides just price of farm product may encourage technology adoption. These incentives can have different forms, such as financial incentives and financial aid by the government (e.g., loan subsidies, tax breaks), support of demonstrator sites to reduce farmer risk, experimental stations or extension services that serve as available consultants to the farmers, a systems approach to integrated weed management, and automation in a systems approach.

8 Future Prospects

The sigmoid curve of technology adoption is also caused by the age of the installed technology that it can replace as concluded by Baerenklau and Knapp (2007). "*A group of identical agents who differ only in terms of the ages of their installed technologies can produce a gradual shift to a new technology with a long-run expected level of adoption that is less than 100 %. Other forms of heterogeneity(e.g., land quality, local markets, producer characteristics, etc.) certainly exist, but neglecting age structure may exaggerate the relative importance of these factors.*" As an example of the introduction of a new platform using technology like weed recognition and weeders, it is a major challenge for the future to build an intra-row weeder that can locate weed plants in a full field-sown crop such as spinach or carrots. Some organic farmers in the Netherlands believe that this is possible. However, the challenge will be to locate the research expertise and entrepreneurs within and outside the agricultural industry to develop new types of weeders into a working system within 10 years or so (Bleeker et al. 2011).

Given all the above considerations, it is not easy to make a forecast for automated weed control adoption. If we take automated mechanical weed control as a first technological innovation, then one can assume that organic agriculture will be the primary target for adoption. At this moment, organic agriculture has a market share of about 5 % (Chap. 2). This is expected to increase to 10 % in the coming decades. Currently, conventional agriculture purchases in Europe around 10,000 new sprayers per year of which 10 % again are sophisticated self-propelled sprayers, probably mainly used by contractors. The sprayers are used for weed control as well as for pest control. Assuming that the ratio of (sophisticated) automated weed control units in organic agricultural to the number of sprayers in conventional agricultural is about the same as the market share of organic agriculture, then the annually newly purchased units of these mechanical systems would be 1,000 units in the EU. This will not happen immediately but might be the case in 20 years. In 10 years that would amount to 500 units per year, a midpoint of a logistic model. The period of 20 years is similar or even a bit less than what is the experience with robotic milking. Lely sold 9,000 units of mil robots over a 20-year period.

Worldwide this number can be double or three times the one estimated for the EU, although it can be argued that the adoption rate in different regions depends on legislative measures, government incentives, and the increase of herbicide resistance. All these may vary by region. Conventional agriculture will also need these mechanical weed control systems, but this will be a slow adoption in the coming 5–10 years. After that, weed resistance to chemical compounds may require a faster adoption. It is more likely that conventional agriculture will first turn to precision automated spraying, which will be primarily a replacement market. This will go slow since at this moment there are a large number of sprayers in use that were purchased recently with, in most cases, nearly state-of-the-art technology. At the same time, crops will be genetically engineered to resist every single chemical herbicide known (the super-stacked), which will eventually lead to every herbicide becoming ineffective on weeds. At this point, targeted identification and alternative (automated) systems for weed control will experience an increase in demand.

9 Conclusions

Technology adoption in agriculture is undefined as to whether it is a new system that must be adopted or whether it can be just a "drop-in" (Schueller, personal communication, 2013). Example of a system change is the introduction of tractors that took a long time for adoption, in part because implements needed to change, horses to get rid of, and farmers needed to get a mechanical aptitude. Precision agriculture is also a new system to be adopted by farmers. Equipment changes as agronomical changes are involved. Adoption of precision agriculture technologies has been fastest where labor is costly but land and capital are relatively less costly. Where precision agriculture is being promoted, the uneven adoption rate is tied to normal cycles for replacing the expensive machinery in which many precision agriculture technologies are embodied (Swinton and Lowenberg-DeBoer 2001). The adoption of hybrid seed corn was technically a "drop-in." Farmers could use the same planter and production system. Auto-guidance is also a "drop-in," since it is relatively easy to just add it on a tractor.

In order to forecast adoption of new technology in agriculture, it is useful to look for historical information on introduction of new technology. There are a number of issues that play a role in the adoption, and it is important to make a list of these as complete as possible. Answers to a number of the questions or "educated guesses" may show the driving forces as well as the constraints for automated weed control systems in agriculture. Based on a first check, it can be expected that automated mechanical weeding will have a place in agriculture and more rapidly so in organic cultivation. It will require an effort from experiment stations and farm advisors to establish convincing demonstrations. Conventional agriculture will follow, but in the near future this will most likely be based on advanced automation for precision chemical weed control. When pressure of herbicide-resistant weeds becomes higher and the number of available chemical compounds reduces, the adoption rate in conventional agriculture will accelerate. The advanced automated weed control using multiple physical as well as chemical means incorporated in one tool will be a part of future weed control. However, the farming community may also reinvent, discover, and adopt alternatives to herbicides by changing the cropping systems to come to an integrated weed management (Llewellyn et al. 2004). However, an early adoption of integrated weed management practices and cropping systems would be beneficial for farmers to proactively address the problem of herbicide-resistant weeds. Automated tools for herbicidal and physical weed management tactics will make such proactive management feasible.

References

Baerenklau KA, Knapp KC (2007) Dynamics of agricultural technology adoption: age structure, reversibility and uncertainty. Am J Agric Econ 89(1):190–201

Barrett SCH (1983) Crop mimicry in weeds. Econ Bot 37:255–282

Bleeker P, Huiting H, van der Weide R (2011) New innovations for intra-row weed control. In: 9th EWRS workshop on physical and cultural weed control 6, Samsun, 28–30 March 2011

Duffy-Martini D, Silberberg E (2004) The diffusion of tractor technology. Ms. http://www.econ.washington.edu/user/silber/The%20Diffusion%20of%20Tractor%20Technology.pdf. Link checked 11 May 2013

Garthwaite D (2002) A survey of current farm sprayer practices in the United Kingdom carried out by CSL. Central Science Laboratory, Sand Hutton

Griliches Z (1960) Hybrid corn and the economics of innovation. Science 132:275–280

Harker KN, O'Donovan JT (2013) Recent weed control, weed management, and integrated weed management. Weed Technol 27:1–11

Lewis JI, Wiser RH (2007) Fostering a renewable energy technology industry: an international comparison of wind industry policy support mechanisms. Energy Policy 35(3):1844–1857

Llewellyn RS, Lindner RK, Pannelland DJ, Powles SB (2004) Grain grower perceptions and use of integrated weed management. Aust J Exp Agric 44:993–1001

Myers J (2009) Preventing death and injury in tractor overturns with roll-over protective structures. http://blogs.cdc.gov/niosh-science-blog/2009/01/rops/. Link checked 11 May 2013

NASS (2008) http://nasdonline.org/static_content/documents/1983/d001892.pdf. Link checked 31 August 2013

NN (2012) Increase in milking automation planned. Agrifuture 12(Spring):17–19

NN (2013) http://www.renewableenergyincentives.co.uk/. Link checked 20 May 2013

Rodriguez F (2012) The realities of robotic milking technology today. The Progressive Dairyman. http://www.progressivedairy.com/index.php?option=com_content&id=9244:the-realities-of-robotic-milking-technology-today&Itemid=75. Link checked 31 August 2013

Sfiligoj E (2012) http://www.croplife.com/article/31932/survey-agco-case-ih-and-john-deere-still-top-ag-equipment-choice-for-retailers. Link checked 20 May 2013

Steele J (2009) Innovation diffusion and travelling waves. In: Shennan S (ed) Pattern and process in cultural evolution. University of California Press, Berkeley, pp 163–174. ISBN 9780520255999

Swinton SM, Lowenberg-DeBoer J (2001) Global adoption of precision agriculture technologies: who, when and why? In: Grenier G, Blackmore S (eds) Third European conference on precision agriculture. Agro Montpellier (ENSAM), Montpellier, pp 557–562

UshaRao K, Kishore VVN (2010) A review of technology diffusion models with special reference to renewable energy technologies. Renew Sustain Energy Rev 14(3):1070–1078

Vanston LK (2008) Practical tips for forecasting new technology adoption. Telektronikk 3/4:179–189

Chapter 14
Automation for Weed Control in Least Developed Countries (LDCs)

Renan Aguero, Noel M. Estwick, and Edgar Gutierrez

Abstract The state of the art in automation for weed control makes it foreseeable that the rate of adoption of these technologies will increase significantly in developed countries within the next decade, as specific applications that are effective and affordable become available. For least developed countries (LDCs), international cooperation must be sought in order to avoid a steep increase in the technology divide. Furthermore, automated weed control could play a significant role in supporting the adoption of weed control strategies that are less damaging to the environment while helping improve food supplies for LDCs. However, since the agricultural systems are rather diverse in these countries, several aid programs will be needed. As these programs develop, stories of success by early adopters could play an important role in promoting widespread use of these technologies among LDCs.

1 Introduction

By 2005, every state in the USA had some land certified for organic farming nearing two million hectares (USDA 2007), a clear indication that this production alternative was consolidating there. However, in least developed countries (LDCs), farming systems tend toward the organic approach out of necessity. Developed or not, for

R. Aguero (✉)
School of Agronomy, University of Costa Rica, San José, Costa Rica
e-mail: renan.aguero@ucr.ac.cr

N.M. Estwick
Cooperative Agricultural Research Center (CARC), Prairie View A&M University, Prairie View, TX, USA
e-mail: nmestwick@pvamu.edu

E. Gutierrez
School of Statistics, University of Costa Rica, San José, Costa Rica
e-mail: edgar.gutierrez@ucr.ac.cr

S.L. Young and F.J. Pierce (eds.), *Automation: The Future of Weed Control in Cropping Systems*, DOI 10.1007/978-94-007-7512-1_14,

organic farming to become more widespread in LDCs, the issue of efficient weed control must be addressed. For centuries, the challenges associated with weed control in LDCs have been met with abundant hand labor. Although these challenges are expected to persist for the foreseeable future, we believe that some progress can be made based on the evolution of automation within the agricultural sectors in each country. Hand labor, as the dominant alternative for weed control, reveals a lot about the social development of a country. At the very least, it gives some indication of the social development of a country's rural sector. Once this moral debate is put behind, the fact is that alternatives to hand labor may be "around the corner."

In a study conducted in the United Kingdom, Sorensen et al. (2005) demonstrated that a weeding robot and integrated band steaming could reduce hand weeding by 85 % in sugar beet and 60 % in carrots. Also, machine vision and RTK GPS guidance systems are potential technologies that can aid in weed detection and identification (Young 2010), and thus it is possible that desired spontaneous species that are not strong competitors to the crop and that overcome the negative interaction with benefits such as hosting pest predators could be left unharmed to grow in isles, within the crops.

According to Seelan et al. (2003), precision agriculture (PA) or precision farming can be defined as "a production system that promotes viable management practices within a field, according to site condition." Variability in PA is both spatial and temporal (Stafford 2000). Stafford also points out that information for the PA system is driven by a number of tools. They include the Global Positioning System (GPS), Geographic Information Systems (GIS), remote sensing, Intelligent Devices and Implements (IDI), and computers.

Although challenges such as the development of information technology (IT) infrastructure negatively influence the adoption of PA in developing countries, it is clearly one of the solutions to improving productivity among smallholder farmers in Africa (Cox 2002). A similar situation most definitely exists in LDCs. The adoption of PA technologies promises both environmental and economic benefits (Stafford 2000). In this chapter we discuss benefits, challenges, and the impact of adopting PA and other forms of automation in developing countries. This will help us make an informed decision as to whether or not automation has a role to play in improving weed control in LDCs.

In this chapter we rank LDCs, using World Bank indexes. We then group countries based on a global index of our own and, from there, we analyze each group of countries potential for a successful adoption of automated weed control techniques in their agricultural systems. This is a rather speculative exercise, but the use of the above-mentioned indexes is an effort to keep our analysis as unbiased as possible. It could very well turn out that in the near future, or maybe even while we write this chapter, that such technologies are already in the hands of early adopters or risk-takers who, once they grasp the knowledge and advantages associated with it, give way to a broader wave of late adopters. The actual worldwide ubiquity of cell phone technology is a good example of how this could occur. We submit that the new generation of smartphones could very well ease adoption of automated technologies at the farm level.

Since this is already a complex speculative analysis, it does not take into account the impact of assisted international programs to promote adoption of these technologies in specific countries. At the least, our effort could render information on where and how that aid could be funneled.

2 Agricultural Systems and Need for Automated Weed Control in LDCs

According to Van de Seeg et al. (2010), "a farming system is defined as a population of individual farm systems that have broadly similar resource bases, enterprise patterns, and household livelihoods and constraints, and for which similar development strategies and interventions would be appropriate." Since the majority of poor people around the world live in rural communities, one solution to eradicating suffering lies in the creation of dynamic communities established upon prosperous farming (FAO and the World Bank 2001). In their quest to reduce poverty through enhanced food security and greater export earnings, LDCs face several internal and external difficulties (FAO n.d.). The FAO (n.d.) paper on globalization, agriculture, and the least developed countries lists low productivity, inflexible production and trade structures, low skill capacity, low life expectancy and educational attainments, poor infrastructure, and deficient institutional and policy frameworks as internal difficulties. It also lists increasing integration of markets resulting from globalization and liberalization as external difficulties. Characterizing farming systems for LDCs reveals that they are diverse with respect to geographical region (Table 14.1).

2.1 Agricultural and Pastoral Production System

The agricultural and pastoral production systems are predominantly cultivated fields and pastures, and are the major sources of food provisions for people in northern Burkina Faso; the major crops are pearl millet and sorghum, while cowpeas, groundnuts and cotton are produced to a lesser extent (Rasmussen et al. 2012).

Automated weed control would be important for the initial stage of crops, what is known as the critical period of competition with weeds. Precision hand pulling and cutting devices would be of particular use to these production systems, and can be coupled with non-tilled sowing techniques to avoid disturbing the weed seed bank. These practices may also prove to be of value for sustained weed control schemes.

Table 14.1 Characterization of farming systems for LDCs

Farming system(s)	Region (countries)
	Sub-Saharan Africa
Integrated farming system	Somalia
Rice and tree crop system	Madagascar
Tree top farming system	Angola
Forest-based farming system	Democratic Republic of the Congo, Equatorial Guinea, Mozambique, Central African Republic, Angola, and Zambia
Highland perennial farming system	Ethiopia, Uganda, Rwanda, and Burundi
Highland temperate-mixed farming system	Ethiopia, Eritrea, Lesotho, and Angola
Root crop farming system	Sierra Leone, Togo, Central African Republic, and Benin
Cereal-root crop mixed farming system	Guinea, Togo, and Benin
Maze mixed farming system	Tanzania, Zambia, Malawi, and Lesotho
Agro-pastoral millet/sorghum farming system	Niger, Somalia, Burkina Faso, and Ethiopia
Pastoral farming system	Mauritania, northern parts of Mali, Niger, Chad, Sudan, Ethiopia, Eritrea, Uganda, and Southern Angola
Sparse (arid) farming system	Sudan, Niger, Chad, Djibouti, and Mauritania
Coastal artisanal fishing farming system	Mozambique, Comoros, Zanzibar, Gambia, Casamance region of Senegal, Guinea-Bissau, Liberia, and Sierra Leone
Urban-based farming system	Expected in all countries listed
	South Asia
Rice farming system	Bangladesh
Coastal artisanal fishing farming system	Maldives, Bangladesh
Rice-wheat farming system	Northeast Bangladesh
Highland mixed farming system	Afghanistan
Tree crop farming system	Upland areas of Bangladesh and Nepal
Pastoral farming system	Afghanistan
Urban-based farming system	Expected in most towns and cities in the region
	East Asia and Pacific
Lowland rice farming system	Myanmar, Cambodia, Laos People's Democratic Republic
Root-tuber farming system	Kiribati, Samoa, Vanuatu, Tuvalu, Timor-Leste, Solomon Islands
Highland extensive mixed farming system	Laos, Northern and Eastern Myanmar
Sparse (forest) farming system	Northern Myanmar
Urban-based farming system	Present in most large towns and cities in the region
	Middle East and North Africa
Highland mixed	Yemen
	Latin America and the Caribbean
Coastal plantation and mixed farming system	Haiti

Source: Farming Systems and Poverty: Improving Farmers' Livelihoods in a Changing World, John Dixon and Aidan Gulliver with David Gibbon, FAO and the World Bank (2001)

2.2 Rice-Wheat Cropping Systems

The rice-wheat cropping systems in South Asia provide grain for approximately 8 % of the world's population (Ladha et al. 2003). These systems are widely found in Nepal and Bangladesh and are characterized by soils that cycle between anaerobic conditions during rice cropping and aerobic conditions during wheat cropping (Alam et al. 2005). A major challenge in countries that have rice-wheat cropping systems is to implement high-yielding production systems that are both profitable and sustainable (Timsina and Connor 2001).

The anaerobic-aerobic cycling of these systems, in itself, provides a formidable tool for weed control. Improvements should be achieved in transplanting machines for the anaerobic conditions under which rice is planted. Robots that identify red rice from white rice, based on subtle differences between the two, would be very important to sustain this production system. Small rice harvesters, with a minimum impact on the soil would also prove beneficial, so that later on, wheat can be sown with minimum tillage planters, and could complete the scheme for an efficient weed control approach.

2.3 Urban-Based Farming System

Urban populations across the globe continue to grow rapidly. Urban-based farming systems are characterized by systems which focus on horticultural and livestock production and have an agricultural population of approximately 40 million (World Bank 2008). Foeken and Mwangi (2000) point out that several of the people who practice urban farming in Nairobi, Kenya, do so because they are unable to obtain formal employment when they migrate to the city.

We believe that a further sophistication of this system might eventually allow some people to evolve into providing landscaping and gardening services within the cities, all of which could be better tapped by adopting modern technologies such as automated weed control. This would allow farmers to avoid the use of herbicides, which, in general, are not welcome in city settings.

2.4 Coastal Plantation and Mixed Farming Systems

Coastal plantation and mixed farming systems can be found in Latin America and the Caribbean and have an agricultural population of approximately 20 million (FAO and World Bank 2001). Mixed farming systems also exist in Asia and Africa. Mixed farming systems are those where crops and animals are integrated and are a major part of small-scale agriculture in Asia (Devendra and Thomas 2002). Romney et al. (2003) argue that mixed farming systems dominate the subhumid, cool highland zones and semiarid zones of sub-Saharan Africa because land is more of a constraint in those highly populous areas than labor. Devendra and Thomas (2002) list a number of benefits that can be derived from mixed farming systems, including the diversifying of risk from single-crop production, the use of labor more efficiently, cash flow for purchasing farm inputs, and value added to crops or their by-products. In addition, Mohammad-Saleem (n.d.) claims that they increase bioenergetic efficiency of agriculture and preserve environmental quality.

3 Capacity for Adoption of Automated Weed Control in LDCs

Technology adoption is a complex multifactor process. In order to allow a quantitative analysis, we include data from the World Bank on different parameters for LDCs, when available. Then, for the sake of discussion and to avoid making a case on a single country, we grouped countries according to those parameters. At one end of this spectrum, we believe that aid programs would be necessary in order to modernize agriculture in general and weed control in particular.

In order to synthesize LDCs situation with respect to their capacity to adopt automated weed control, it was decided to summarize their status by using a set of data from the World Bank open database (http://data.worldbank.org/). Four parameters were proposed as a means of portraying the situation surrounding 50 LDCs with respect to their ability to adopt new technologies. They are social setting, schooling setting, knowledge base, and window of opportunity. Data for LDCs in relevant areas were scarce; thus, 21 indicators were chosen and grouped into these four dimensions, some as proxies (Table 14.2).

Data for parameters (Table 14.2) was gathered for the years 2000 to 2010. However, the year 2010 was dropped for lack of data for most of the indicators. Indicator values were rescaled so that all of them ranged from 0 (less desirable) to 1 (more desirable) and thus could be compared and synthetized in indices. For indicators *Poverty gap at rural poverty line* and *vulnerable employment* for female and male, a low value of the indicator was desirable as the opposite for the other indicators. To have the same direction of the others, the rescaling of this indicator was done in such a way that low values of the indicator resulted in high values after rescaling. In this way, all of the indicators were set to aim at high values as desirable targets.

From the set of values made by all data available for years 2000 to 2010, and for the 50 countries that comprise the LDCs, the maximum and minimum values were obtained to rescale them as values from 0 to 1. The formula used was the following:

When high values are desirable, then one aims at 1 as the best value by transforming values using

$$\frac{y_{ij} - y_{i,\max}}{y_{i,\max} - y_{i,\min}}.$$

Table 14.2 Parameters for LDCs to adopt automated weed control technologies

Dimension	Indicator	Value Max	Min
Social setting	GDP per capita (current US$)	27816.2	86.8
	Poverty gap at rural poverty line (%)	53	7.6
	CPIA public sector management and institutions cluster average (1 = low to 6 = high)	4	2.2
	Access to electricity (% of population) – 2009	55	9
	Agriculture, value added (% of GDP)	75.5	2.0
	Employment in agriculture (% of total employment)	85.1	2.8
	Income share held by lowest 20 %	9.36	2
Schooling setting	Literacy rate, youth total (% of people ages 15–24)	99.5	14.0
	Public spending on education, total (% of GDP)	16.8	0.6
	School enrollment, primary (% gross)	156.3	19.3
	Trained teachers in primary education (% of total teachers)	100	14.6
	Agriculture value added per worker (constant 2000 US$)	3311.6	65.4
Knowledge base	Domestic credit provided by banking sector (% of GDP)	235.1	−31.8
	Researchers in R&D (per million people)	275.5	6.2
	Research and development expenditure (% of GDP)	0.53	0.006
	Technicians in R&D (per million people)	137.2	2.5
Window of opportunity	Life expectancy at birth, total (years)	76.2	39.7
	Roads, paved (% of total roads)	100	0.8
	Income share held by lowest 10 %	4.3	0.6
	Vulnerable employment, female (% of female employment)	95.9	1.7
	Vulnerable employment, male (% of male employment)	89.3	2.1

When low values are desirable, then, to have the same direction as the others, one aims at 1 as the best value by transforming values using

$$\frac{y_{\max} - y_{ij}}{y_{\max} - y_{\min}}$$

where y_i is the reported value for the ith indicator and jth year and $y_{i,\max}$ and $y_{i,\min}$ are the maximum and minimum values for the ith indicators for the years 2000 to 2010. Annex A shows the data for the indicators used and the transformed values as well.

This exercise takes one to the comparison of the performance of LDCs with respect to the four dimensions. Keeping in mind that 1 is the desirable value, and using only the LDCs for the comparison, then such performance is portrayed (Fig. 14.1). The comparison is made among LDCs (i.e., the reference value for an indicator is the best value observed for the 50 countries during the study period). All dimensions

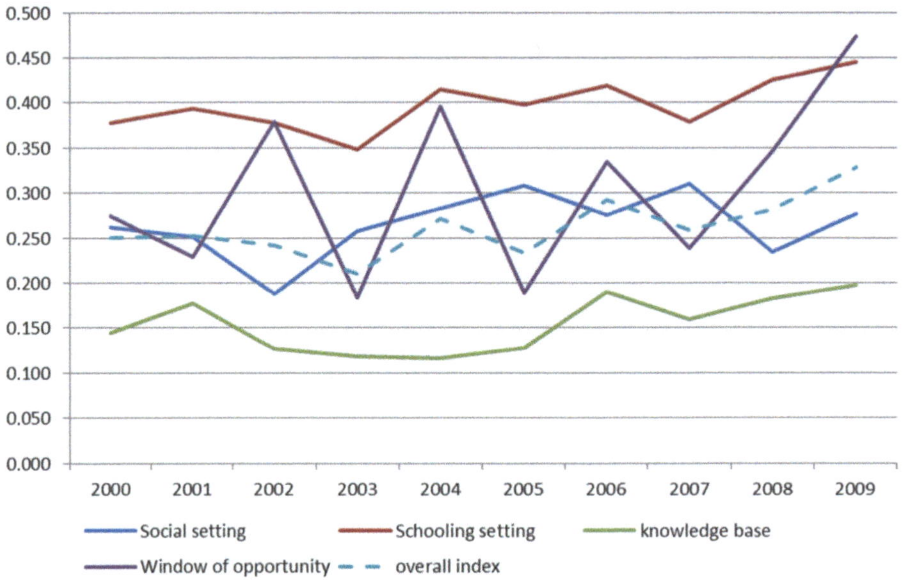

Fig. 14.1 LDCs performance in social setting, schooling setting, Knowledge base and window of opportunity for a successful adoption of automated weed control techniques. 2000–2009 (Source: World Bank data set (http://data.worldbank.org/) consulted 17 Jan 2012)

resulted in values below 0.5 insinuating low performance, for the adoption of new weed control techniques. *Knowledge base* and *Window of opportunity* are the two dimensions which present more limitations to achieve a good scenario for the adoption of new techniques. Knowledge base because during the study period it keeps very low values, all below the 0.2 level, meaning that having as the reference the best values for LDCs, yet good values are not shown when the group of countries is considered as such. As a result of its erratic behavior, the *Window of opportunity* dimension also shows a limitation for the adoption of new weed control techniques because it would be very difficult to introduce a systematic program for introducing new technologies. Despite the fact that *Schooling setting* shows the highest values of the four dimensions, great variability is also observed among countries. This is a common characteristic for the set of dimensions.

Analyzing the pattern of the *Overall index* which depicts the combined pattern of the four dimensions in (Fig. 14.1), clearly demonstrates that the performance of LDCs in this context is poor. The overall score reaches just 25 % of the full score that can be obtained using the best values of countries found within LDCs and the studied period of time.

The way the four dimensions behaved for LDCs during the chosen years is quite different (Fig. 14.2). Here it is shown how *Social setting* and *Knowledge base* have maintained a very stable pattern, whereas *Schooling setting* and *Window of opportunity* are the two that show increments during those years; however, it has to be

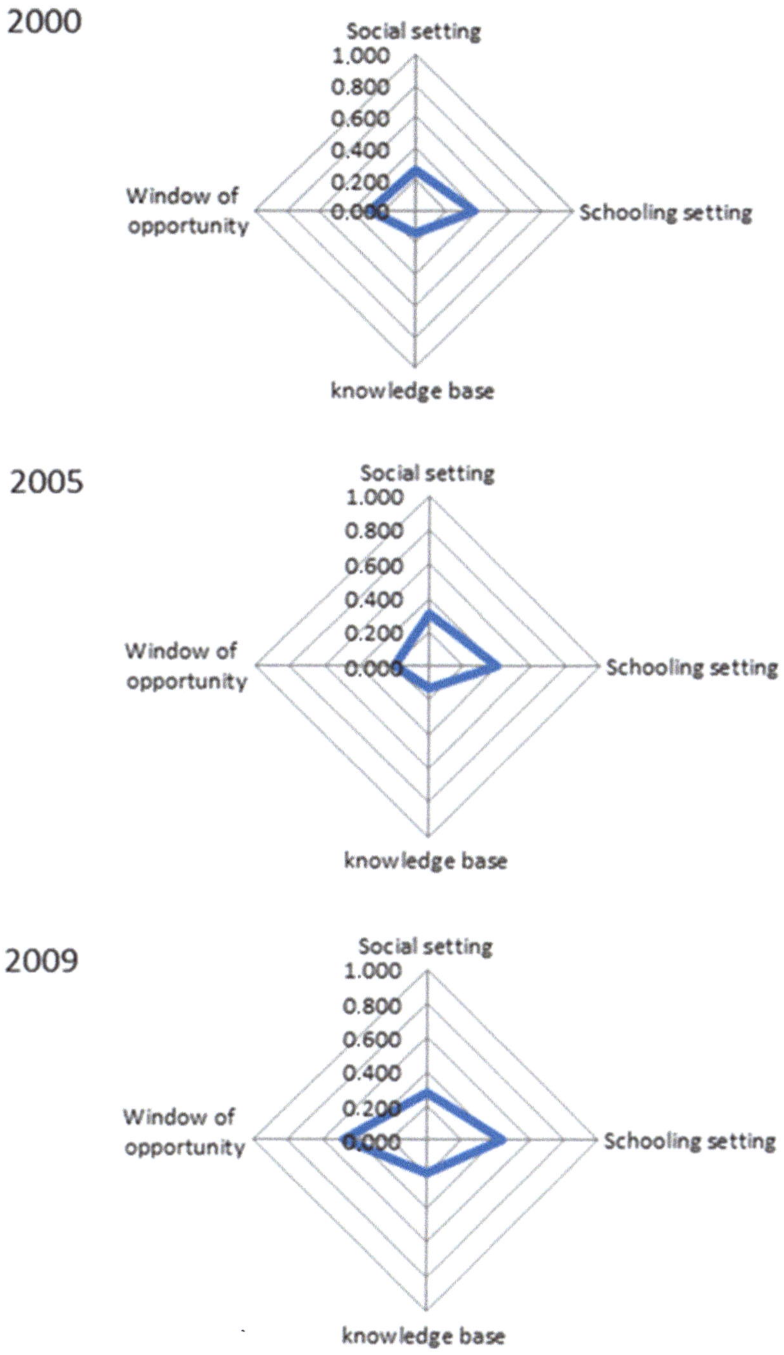

Fig. 14.2 LDCs performance in social setting, schooling setting, Knowledge base and window of opportunity for a successful adoption of automated weed control techniques. 2000, 2005, and 2009 (Source: World Bank data set (http://data.worldbank.org/) consulted 17 Jan 2012)

noted that the heterogeneity within countries for those dimensions is very ample. Beyond the low values shown by these dimensions, the most outstanding characteristic of how LDCs behave with respect to these indicators is the wide variability of results in each of them. This demonstrates the great differences that exist within countries in spite of being part of the same group.

4 Potential for Adoption of PA and Automation in LDCs

Maohua (2001) observed that poverty and agriculture in developing countries are intertwined, and added that traditional farming persists with limited application of modern technology. Despite the fact that little automation exists in the agriculture sector in developing countries, the future appears to be promising. Since worse conditions exist in LDCs, adopting automation is a daunting task. Kassler (2001) commented that the forecast of processes becoming increasingly influenced by automation in years to come is a credible consequence of the anticipated advancements in science. Maohua (2001) appears to support this by claiming that new opportunities for the transformation of traditional farming into modern agriculture will be created as a result of the information and knowledge era. Even though resource-poor countries will not be at the forefront of advancements such as automated weed control, we are optimistic that with careful planning, policy development, and implementation of technologies and innovations, LDCs can improve their agricultural production systems. For example, the Consultative Group on International Agricultural Research (CGIAR) has taken steps in this direction by adopting sustainable intensification as one of its initiatives in the MAIZE program (CGIAR 2011).

Stafford (2000) identified four drivers that will provide a stimulus for the implementation of PA in the twenty-first century; they are environmental legislation, traceability, public concern over farm practices, and the public dislike for genetically modified crops. Further analysis of these drivers will demonstrate how they impact the adoption of PA in developing countries. Maohua (2001) suggests that developing countries can improve their chances of adopting PA by enhancing their information infrastructures, thereby decreasing the poverty of information and related imbalance in socioeconomic development. LDCs would benefit from similar strategies if they could attract the necessary resources.

The theory of Diffusion of Innovations may play a key role in understanding how PA is adopted in LDCs. Rogers (2003) divided adopters into categories based on innovativeness. The categories are innovators, early adopters, early majority, late majority, and laggards. Rogers (2003) argues that their percentages of adoption are approximately 3, 13, 34, 34, and 16 %, respectively. Following is how Lamb et al. (2008) explain the categories. Innovators comprise the technocratic minority who have greater access to information, are most proactive, and consequently most likely to be the risk-takers in technology adoption. A larger group of early adopters who are leading farmers follow the innovators. We expect research to support the presence of these two groups in LDCs.

5 Conclusions

The rate at which developing countries can obtain the technology infrastructure, GPS, GIS, and RS will prove to be a significant limitation on their adoption of agricultural innovation. Enhancing information infrastructures to reduce the poverty of information and related imbalance in socioeconomic development will also prove pivotal in determining the rate at which PA can improve agricultural production. Analysis of World Bank data for LDCs over the period 2000 to 2010 revealed the following:

1. Recent data does not provide a positive picture for adoption of automated weed control in LDCs.
2. If assisted, there are many opportunities for adoption of automated weed control to solve many weed control challenges among the diverse production systems present in the LDCs.
3. Early adopters and risk-takers may hold the keys to influencing the adoption of automation technologies in LDCs.

References

Alam MM, Ladha JK, Foyjunnessa Z, Rahman SR, Khan H-u-R, Khan AH, Buresh RJ (2005) Nutrient management for increased productivity of rice-wheat cropping system in Bangladesh. Field Crop Res 96:374–386

CGIAR (2011) MAZE – global alliance for improving food security and the livelihoods of the resource-poor in the developing world. Retrieved from http://consortium.cgiar.org/wp-content/uploads/2011/10/Maize-flyer-light.pdf. On 18 Jan 2012

Cox S (2002) Information technology: the global key to precision agriculture and sustainability. Comput Electron Agric 36(2):93–111

Devendra C, Thomas D (2002) Smallholder farming systems in Asia. Agric Syst 71:17–25

FAO (n.d) Globalization, agriculture and the least developed countries. Retrieved from http://www.unohrlls.org/UserFiles/File/LDC%20Documents/Turkey/20June07-Agriculture-Final.pdf. On 12 June 2012

FAO and the World Bank (2001) Farming systems and poverty: improving farmers' livelihoods in a changing world. Retrieved from ftp://ftp.fao.org/docrep/fao/003/y1860e/y1860e.pdf. On 21 June 2012

Foeken D, Mwangi AM (2000) Increasing food security through urban farming in Nairobi. In: Bakker N, Dubbeling M, Güendel S, Sabel-Koschella U, de Zeeuw H (eds) Growing cities, growing food. DSE, Feldafing, pp 303–328

Kassler M (2001) Agricultural automation in the new millennium. Comput Electron Agric 30:237–240

Ladha JK, Dawe D, Pathak H, Padre AT, Yadav RL, Singh B, Yadvinder Singh Y, Singh P, Singh AL, Kundu R, Sakal N, Ram AP, Regmi SK, Gami AL, Bhandari R, Amin CR, Yadav EM, Bhattarai S, Das HP, Aggarwal RK, Gupta RK, Hobbs PR (2003) How extensive are yield declines in long-term rice-wheat experiments in Asia? Field Crop Res 81:159–180

Lamb DW, Frazier P, Adams P (2008) Improving pathways to adoption: putting the right P's in precision agriculture. Comput Electron Agric 61:4–9

Maohua W (2001) Possible adoption of precision agriculture for developing countries at the threshold of the new millennium. Comput Electron Agric 30:45–50
Mohammed-Saleem MA (n.d.) Mixed farming systems in sub-Saharan Africa. Retrieved from www.fao.org/wairdocs/ilri/x5462e/x5462e0e.htm. On 7 Aug 2012
Rasmussen LV, Rasmussen K, Reenberg A, Proud S (2012) A system dynamics approach to land use changes in agro-pastoral systems on the desert margins of Sahel. Agr Syst 107:56–64
Rogers EM (2003) Diffusion of innovations, 5th edn. Free Press, New York
Romney DL, Thorne P, Lukuyu B, Thornton PK (2003) Maize as food and feed in intensive smallholder systems: management options for improved integration in mixed farming systems of east and southern Africa. Field Crop Res 84:159–168
Seelan SK, Laguette S, Casady GM, Seielstad GA (2003) Remote sensing applications for precision agriculture: a learning community approach. Remote Sens Environ 88(1):157–169
Sorensen CG, Madsen NA, Jacobsen BH (2005) Organic farming scenarios: operational analysis and costs of implementing innovative technologies. Biosyst Eng 91(2):127–137. doi:10.1016/j.biostemseng.2005.03.06
Stafford JV (2000) Implementing precision agriculture in the 21st century. J Agric Eng Res 76:267–275. doi:10.1006/jaer.2000.0577
Timsina J, Connor DJ (2001) Productivity and management of rice-wheat cropping systems: issues and challenges. Field Crop Res 69:93–132
USDA (2007) The census of agriculture United States Department of Agriculture. Retrieved from http://www.agcensus.usda.gov/Publications/2007/Full_Report/usv1.pdf. On 5 Jan 2012
Van de Seeg JA, Verburg PH, Baltenweck I, Staal SJ (2010) Characterization of the spatial distribution of farming systems in the Kenyan Highlands. Appl Geogr 30:239–253
World Bank (2008) Managing land and landscapes. Retrieved from http://siteresources.worldbank.org/INTARD/927371-1205790395237/21689740/SourcebookMarch08.pdf. On 7 Aug 2012
Young SL (2010) Weed control in organic cropping systems. Resour Mag 17(1):8–9

Section VI
Future Directions

Chapter 15
Future Directions for Automated Weed Management in Precision Agriculture

Stephen L. Young, George E. Meyer, and Wayne E. Woldt

Abstract In cropping systems, integrated weed management is based on diversification. Rather than relying solely on one or two herbicides, a multiplicity of weed control strategies is employed. Yet, integrated weed management as currently practiced is far from integrated; every weed is still managed the same regardless of location or season. The recent development of precision application technology is now allowing for smaller treatment units by making applications according to site-specific demands. The automated systems of the future will have sensor and computer technologies that first categorize each and every plant in the field as either weed or crop and then identify the species of weed. Following identification, multiple weed control tools located on a single platform are applied at micro-rates to individual plants based on their biology. For example, if the system identified a weed resistant to Roundup, it could be spritzed with a different herbicide or nipped with an onboard cutter or singed with a burst of flame. This system and others like it will be capable of targeting different weed-killing tools to specific weeds. This chapter will discuss the challenges and tools of the future.

S.L. Young (✉)
Department of Agronomy and Horticulture, West Central Research & Extension Center, University of Nebraska-Lincoln, 402 West State Farm Road, North Platte, NE 69101, USA
e-mail: steve.young@unl.edu

G.E. Meyer
Biological Systems Engineering, University of Nebraska-Lincoln, 244 L.W.C., Lincoln, NE 68583, USA
e-mail: gmeyer1@unl.edu

W.E. Woldt
Biological Systems Engineering, University of Nebraska-Lincoln, 232 L.W.C., Lincoln, NE 68583, USA
e-mail: wwoldt1@unl.edu

S.L. Young and F.J. Pierce (eds.), *Automation: The Future of Weed Control in Cropping Systems*, DOI 10.1007/978-94-007-7512-1_15,

1 Introduction

The population of the world has surpassed seven billion and is expected to reach nine billion by 2050. This presents a global challenge involving the land and resources available. Current estimates indicate 1.2 acres are required to feed a single person (Giampietro and Pimentel 1994). The total land mass of the world is approximately 149 million km^2 where 12–18 % is arable land suitable for crop production. If it takes 1.2 acres to feed one person and there are only 27 million km^2 of arable land, we can only feed 5.5 billion people, which is dreadfully short both now and into the future. From a pragmatic perspective, the conceptual options are reduce the population of the world, increase the amount of arable land, or increase the production efficiency of the number of acres to feed one person. It is obvious that the first two are not likely to occur, so we are left with the last option. To accomplish this in a reasonable period of time will require new planning, funding, research, and outreach. Possible solutions include advances in weed management, improved efficiency in irrigation, progress in genetic research, and the development of better precision crop management. Precision in crop management requires knowing more about plant, soil, and climatic conditions and how to adjust and accommodate changing soil and environmental conditions.

Producers are often faced with challenges from both the environment and the application of technology to various scales and complexity of production that exists locally, nationally, and even globally. In 2012, several regions across the globe (e.g., central USA) were in a drought resulting in extremely challenging conditions for successfully growing crops. Irrigated agriculture did well, while dry land producers realized significant negative economic impacts. Producers also faced challenges with high input costs, such as fuel prices to run tractors and machines, nutrient expenses to improve yield, and pesticide costs to reduce losses, not to mention the associated societal impact.

In the future, it is highly probable that commercial fertilizers (e.g., phosphorous, nitrogen) and water for growing crops will not be as readily available, and, therefore, we will be challenged with how to adequately supply our crops with fertilizers and water. These challenges can be captured under the broad concept of a yield gap (Lobell et al. 2009). The yield gap can be defined as the difference between yield potential that could be achieved under ideal production and the yield obtained under current production. Closing this yield gap will contribute to meeting the food and fiber demand for our increasing global population.

As noted earlier, improved weed management and precision agriculture have the potential to contribute to solving the yield gap challenge. Weeds compete with crops for light, nutrients, and water. Weedy and invasive plants cost the world economy billions of dollars annually in crop damage and lost earnings. In the USA, various states have reported annual weed control costs in the hundreds of millions of dollars. Herbicides account for more than 72 % of all pesticides used on agricultural crops. Furthermore, it is estimated that $4 billion was spent on herbicides in the USA in 2006 and 2007 (Grube et al. 2011).

With the increased use of precision crop management comes the need for gathering crop, soil, and environmental information and implementing machine control. The typical paradigm for crop management in Controlled Environment Agriculture (CEA) is to focus on the status and health of individual crop plants. In many ways, the health of field-grown plants is affected by weeds. For example, each individual weed in a field consumes water and nutrients that could be used by crops. Numerous other considerations must be accounted for, and we can now develop the technology to precisely address individual plants, both crop and weed. Unfortunately, the wind, rain, and environmental elements create difficult conditions for easily and quickly making targeted treatments to individual leaf surfaces or small plants. In addition, terrain and spatial distribution of crops and weeds can be nonuniform and difficult conditions to address that are independent of weather and climate. These challenges currently do not have a simple answer, especially with limited funding for perceived high-risk research projects. Therefore, the perception of national and international agricultural policy managers, many in industry, and financial investors that control investment capital needs to change if solutions to the limited world food supply are to be obtained.

2 Future Patterns

Agricultural land use dropped slightly from 54 to 51 % between 1982 and 2007, labor input declined 30 %, but productivity increased 50 % (O'Donoghue et al. 2011). During the same period, adoption of new technologies increased dramatically. Plant and soil sensors have been a rapidly developing area of technology with widespread adoption in many fields, including agriculture. From Global Positioning Systems (GPS), to guidance, to the potential use of robots for weed management, agriculture has advanced rapidly in recent decades.

In the health and environmental sciences, recent developments have included new sensors at the microscale. At Georgia Tech, scientists are inserting nanopiezo-electronics into the human body to detect early signs of disease in blood, detect minute amounts of poisonous gases in air, and to find trace contaminants in food. These devices are very sensitive, run on low power, some from minuscule generators, but tiny in size. A startup laboratory, BioNanomatrix (now BioNano Genomics), is pursuing the key to personalized medicine, which is based on the rapid computer assessment that can sequence an entire genome in 8 h for a mere $100. With this powerful tool, medical treatment could be tailored to a patient's distinct genetic profile. Perhaps a similar approach can be applied to signature plants within a given field, as an early warning system for biological stress.

Other available or developing technologies that use sensors and embedded computing systems are pill cameras that are remote controlled for movement within the digestive system with muscular contractions. Smaller but still relatively high-resolution cameras result in lighter payloads and smaller energy requirement for robots and other deployment systems to gather plant information. Another type of

Fig. 15.1 The light field camera (Drawing Courtesy of S.A. Smith, Graphics Artist, Biological Systems Engineering, University of Nebraska)

camera, which was proposed back in 1908 and is receiving recent attention, is the light-field camera (Harris 2012) (Fig. 15.1). Also called a plenoptic camera, it uses a microlens array to capture 4D light-field information about a scene. The plenoptic camera features a matrix of tiny lenses on a sensing chip. These sensors gather light from different sources and directions. Such light-field information can be used to develop a three-dimensional database of the features of complex scenes, such as a canopy. Changing the focal plane would allow one to look deep inside canopies for pests and disease on leaves that are exposed to the sky.

It is obvious that the trends are moving society toward more integration with technology. In cropping systems, a combination of biology and engineering has recently merged to address management tools designed to respond to the dynamics of nature in the land, air, and water.

3 The Need for Change

Current weed control practices lack the precision needed to effectively and safely control weeds without harmful side effects. Organic and conventional producers rank weed control as their number one production cost. For organic producers particularly, weed control has become increasingly important as organic production has increased its market share. In conventional systems, herbicide resistance, off-target movement, and increased regulations have left many growers with few alternatives. Added to this is an increasing demand from the public for a safer and more sustainable supply of food (see Chap. 2). The problems of current mechanized agricultural systems have set the stage for the introduction and adoption of more advanced technology to meet the needs of growers and satisfy the desires of consumers.

Automation and sensor technology continues to expand rapidly with advancements in all fields, including medical, mechanical, and analytical sciences (see Chap. 3). The applications to agriculture have occurred at a slower pace, but new technology is changing this. For the weed scientist, plant biology is one of the most important factors for developing weed control strategies. Without an understanding of the changes that occur during plant growth and development, most weed control practices will be less than satisfactory. Critical to automated control will be an account of weed morphology and the precise periods of when control measures need to be applied (see Chap. 4).

Primarily, weeds are controlled mechanically or chemically in current cropping systems. Mechanical disturbance or destruction is applied to weeds with blades, bars, discs, or other steel instruments that move at a continuous pace through the field following a designated path (see Chap. 7). Similarly, herbicide applications are made indiscriminately to plants, either weeds (e.g., directed spray) or crops and weeds (e.g., herbicide-resistant crops), using broadcast spray equipment (see Chap. 8). Both mechanical and chemical methods of weed control make inputs based on the general or average condition of the field without accounting for the spatial and temporal changes that occur at microscales.

In some parts of the world, there is a pressing need for more precise weed control using advanced technology (see Chap. 10), while in other regions, regulations are less stringent and development is occurring only for select and small market crops (see Chap. 9). Globally, economics are the biggest driver for the adoption of precision weed management technologies (see Chaps. 12 and 13). Even in least developed countries, the use of technology for precision weed control has potential but not without support from government programs and cooperation with other nations (see Chap. 14).

4 What Lies Ahead?

Production agriculture is contributing in meeting the needs of a growing population, but our methods for growing food must get better faster or we could face a significant shortfall. One way to do this is by being more precise in our management of pests (e.g., weeds), which will result in increased production, lowered inputs, and reduced environmental contamination, which in many ways moves us closer to more sustainable systems.

Precision weed management (PWM), which simply stated "places the right amount of inputs on the right target [weeds] at the right time," is an approach to managing weeds that is better for the environment and better for the producer as it leads to a reduction of inputs without decreasing weed control efficacy. In fact, one of the biggest contributions of PWM is the improved efficacy of controlling virtually all weeds in any cropping system (e.g., conventional, organic) (Fig. 15.2). This shift in approach is based on strong collaborations between biologists, computer scientists, and engineers who are working to harness tools with powerful technology and use them to better manage weeds, which are a major problem in cropping systems throughout the world.

Fig. 15.2 Weed robots of the future will have all of the control tools and a decision support system on a single platform that moves autonomously through the field. Not only that, but also UASs will circle overhead and work directly with on-the-ground robots through wireless communications (Drawing courtesy of S.L. Young)

4.1 Plant Recognition

Most studies during the last 20 years have addressed the classification of only two crop-weed classes or general cases of broad leaf versus grasses and in other cases, crop row versus between the crop row (Tang et al. 2003). However, to precisely classify a plant species that may be imbedded within other different species of plants in an image is a botanically challenging exercise. Due in large part to advanced sensor and computing technology, it is now possible to put together a complete robust system that essentially mimics the human taxonomic, plant identification keying method. Future studies are needed to determine minimal digital image resolutions needed to maintain the highest species discrimination performance.

Fuzzy logic, cluster algorithms, and cluster reassembly routines mimic human perception and decision-making and tend to work well for extracting convex leaf shapes from plant canopy images (Neto et al. 2006). However, for more botanically diverse leaf shapes, such as species with complex leaves, lobed margins (indented), and trifoliolates, new fitness criteria must be developed to accommodate various leaf shapes. Undoubtedly, integration of specific shape and textural venation feature analyses as a fitness or classification criteria may be a key to improvement for plant species identification (Price et al. 2011). Work has already begun on utilizing digital canopy architecture metrics such as three dimensions, which is important to plant taxonomy.

Table 15.1 UAS examples and application areas focused on weed management

Examples	Application areas
Precise placement of optical and thermal sensors	Weed control applications
Sensors for natural resource management	Invasive species detection/mapping
Crop scouting opportunities	Infestation detection/mapping
Soil moisture and vegetation type/index	Crop canopy condition/growth stage
Pesticide management and field application	Space/time resolution; crop dusting
Remote sensing with multispectral sensors	FLIR, LIDAR, gas/moisture flux

4.2 Aerial Technology

Collection of very detailed plant canopy and soils production information may be implemented using Unattended Aerial Systems (UASs) equipped with various sensors and extended local wireless data collection capabilities. The UAS offers a unique opportunity to place crop and soil sensors, robotics, and advanced information systems at more timely and desired field locations for increasing production and improving efficiency of agricultural operations. The opening of National Airspace to UAS, currently scheduled for the fall of 2015, has the potential to make a significant contribution to closing the yield gap and could be a "game changer" for the agricultural industry. The potential application of UAS in agricultural and natural resources are wide and varied (Table 15.1).

The effectiveness of an aerial system as such will depend on its ability to both cover a large expanse of cropland and levitate or focus over desired areas of the crop field. In addition, the UAS will need to carry an appropriate sensor and data storage package, remain on a given target area, and work in tandem with on-the-ground robots via wireless communications. Remote sensing has been traditionally available from satellites and low-flying manned aircraft in the past, however, with variable data quality and high cost. Current trends are that fixed wing UASs have the longest range with flight time measured in miles and hours, compared to multi-propeller helicopters, some with only 20 min of air time. Terrestrial robot systems, especially small ones, may also have short operation times due to energy demands and our current state of energy system density, with lithium polymer batteries being the state of the art. Electro-optical (photonic) sensor technology is quite advanced. Yet another extension of these thoughts includes the concept of a UAS collaboration network, in which multiple UASs work together (perhaps even employ swarm technology), along with an array of terrestrial-based robots, to realize an integrated, collaborative framework that achieves outcomes that are not possible by single robotic systems. The UAS industry has identified precision agricultural applications, including weed control and management, as the single largest market opportunity through year 2025 (Jenkins and Vasigh 2013). Research and development on the deployment of information gathering and subsequent control technology is the current limiting factor (Fig. 15.3).

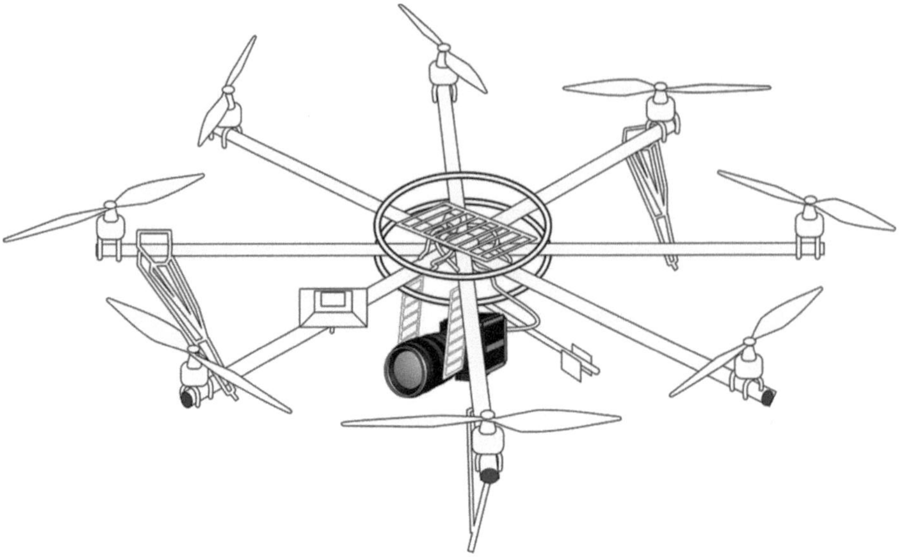

Fig. 15.3 Multi-rotor Unmanned Aircraft System (UAS) (Drawing courtesy of S.A. Smith, Graphics Artist, Biological Systems Engineering, University of Nebraska)

5 What Do We Need to Do?

Success on PWM is based on the integration of expertise from multiple fields of study that can address a problem that has plagued agriculture from its very start: weeds. Even before the introduction of the first herbicides, researchers had been developing biological methods and engineering approaches to control weeds. Since then and more recently, a reliance on herbicides eliminated the need for real advancement in weed management, and subsequently engineers and biologists tended to work separately.

Today, the broadcast application of herbicides is impacting our ecosystems (e.g., runoff, drift, ground water contamination) and causing entire cropping systems to fail (e.g., herbicide-resistant weeds), signaling the need for renewed collaboration between biologists and engineers. Considering the increasing number of people on this planet and the little amount of time to reconcile how to feed them all, we cannot afford to have our current systems fail, let alone ignore what is needed for the future.

In an effort to address this need, a paradigm shift is needed by those involved in weed control in cropping systems from the grower to the consultant to the researcher. If we expect to continue to maintain current yields and also increase production in the future, we will have to think more broadly in incorporating alternative approaches in our management strategies. A possible starting point is the model by Zijlstra et al. (2011) for a crop protection system of the future that is based on novel

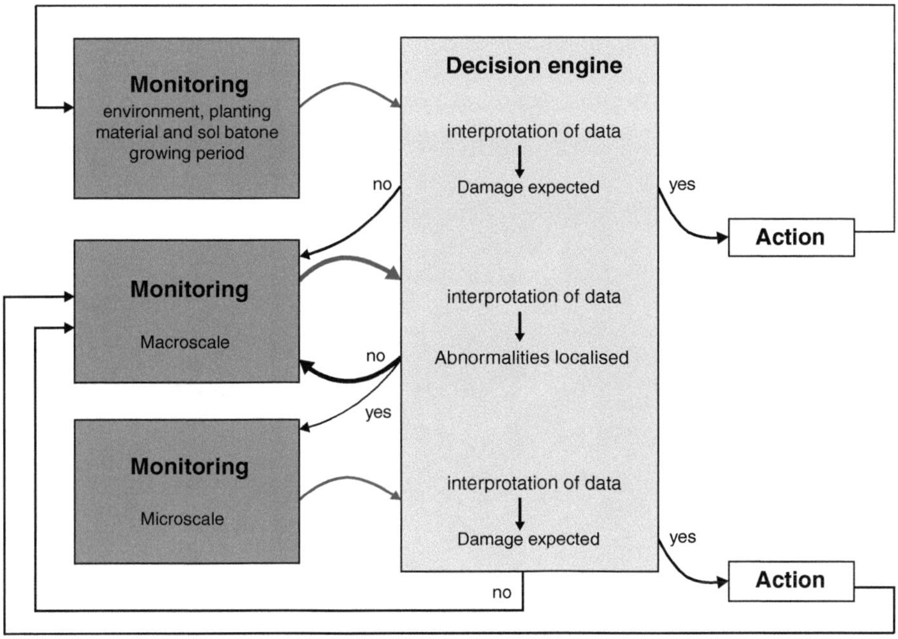

Fig. 15.4 Generic model of an innovative crop protection system. The first monitoring step and the subsequent decision step are performed before the growing period; the other steps are performed during the growing period (Reprinted with permission from Zijlstra et al. 2011)

monitoring tools and precision application technologies (Fig. 15.4). With a historical account of the field and following the planting of the crop, monitoring is conducted on macro- and microscales. At any point, certain locations or "hotspots" could be targeted in the field with the appropriate and precise action made at individual plant scales. According to Zijlstra et al. (2011), the two-step monitoring from macro- to microscale levels would enable earlier detection of pests (e.g., weeds) and thus require new action thresholds and dose-response relationships, which would require substantial research effort.

The ability to monitor an entire field at the individual plant scale will require the use of communications, mobile devices, and decision support software, which make up swarm technology. In the future, the use of robots working in a coordinated effort to manage individual crop plants or swarm through a field may become more important than genetically engineered seeds or new fertilizer formulations. Dorhout R&D LLC (http://dorhoutrd.com/home/prospero_robot_farmer) has developed a small six-legged robot ("Prospero") that has successfully planted an Iowa cornfield in a test run (Fig. 15.5). Robots can make very precise decisions about where and when to plant seeds based on different kinds of soil type within the same field. Rather than using a GPS device to lay a precise line of seeds in a field, the Prospero robots talk to each other as they crawl, staying within about 2 meters of each other. Eliminating a GPS unit helps keep the robots "brains" simple as well as lowering the cost of each unit.

Fig. 15.5 Prospero, the robot that plants and manages crops fields of the future (Reprinted with permission from Dorhout R&D LLC (LLC (http://dorhoutrd.com))

The next step for the small company is scaling up the robots and, as with all autonomous devices, pushing battery power to extend their daily life span. Given the increasing demands on farms to produce more food, and the small margins on which farmers operate, Dorhout R&D LLC and others expect to see more automation in the fields in the near future.

It is safe to say that if we could manage weeds without inputting toxins, causing erosion, and changing genetics, we would. Unfortunately, the population of the world is increasing, yet the amount of arable land available for producing crops is not. Therefore, we need to get more precise in managing crop production and at the same time take steps to protect and limit damage to the ecosystems that ultimately support all life forms in all parts of the globe.

References

Giampietro M, Pimentel D (1994) Food, land, population, and the U.S. economy. Carrying Capacity Network, Washington, DC

Grube J, Donaldson D, Kiely T, Wu L (2011) Pesticides industry sales and usage – 2006 and 2007 market estimates. US Environmental Protection Agency, EPA 733-R-11-001. http://www.epa.gov/pesticides/pestsales/07pestsales/market_estimates2007.pdf

Harris M (2012) Light-field photography revolutionizes Imaging. IEEE Spectrum. May. http://spectrum.ieee.org/consumer-electronics/gadgets/lightfield-photography-revolutionizes-imaging/0

Jenkins D, Vasigh B (2013) The economic impact of unmanned aircraft systems integration in the United States. Association for Unmanned Vehicle Systems International, Arlington, 38 pp
Lobell DB, Cassman KG, Field CB (2009) Crop yield gaps: their importance, magnitudes, and causes. Annu Rev Environ Resour 334:1–26
Neto JC, Meyer GE et al (2006) Individual leaf extractions from young canopy images using Gustafson-Kessel clustering and a genetic algorithm. Comput Electron Agric 51:65–85
O'Donoghue E, Hoppe RA et al (2011) The changing organization of U.S. farming. USDA Economic Research Service, EIB No. 88. http://www.ers.usda.gov/media/176816/eib88_1_.pdf. Accessed 8 Oct 2012
Price CA, Symonova O, Mileyko Y, Hillery T, Weitz JS (2011) Leaf extraction and analysis framework graphical user interface: segmenting and analyzing the structure of leaf veins and areoles. Plant Physiol 155:236–245
Tang L, Tian L et al (2003) Classification of broadleaf and grass weeds using Gabor wavelets and an artificial neural network. Trans ASAE 46:1247–1254
Zijlstra C, Lund I et al (2011) Combining novel monitoring tools and precision application technologies for integrated high-tech crop protection in the future. Pest Manag Sci 67:616–625

Index

S.L. Young and F.J. Pierce (eds.), *Automation: The Future of Weed Control in Cropping Systems*, DOI 10.1007/978-94-007-7512-1,

MIX
Papier aus verantwortungsvollen Quellen
Paper from responsible sources
FSC® C105338

If you have any concerns about our products,
you can contact us on
ProductSafety@springernature.com

In case Publisher is established outside the EU,
the EU authorized representative is:
Springer Nature Customer Service Center GmbH
Europaplatz 3, 69115 Heidelberg, Germany

Printed by Libri Plureos GmbH
in Hamburg, Germany